This book is to be returned on
or before the date stamped below

CANCELLE

20 JUN 2002

27 JUN 2002

22 MAR 2001

16 JUN 2003

15/6/2001

29 MAR 2004

- 5 JUN 2002

01 JUN 2004

30 SEP 2004

COUNTRYSIDE COMMISSION

The Coastal Heritage

A conservation policy for coasts of high quality scenery

LONDON
HER MAJESTY'S STATIONERY OFFICE: 1970

First published 1970
Second impression 1971

SBN 11 700486 3*

CONTENTS

MAPS

In rear pocket:

Following text (for use with Appendix 2):

PLATES

The plates include representative aerial views of each of the 34 potential Heritage Coasts referred to in Chapter 4. The cover photograph shows Pendower Cove and Nanjizal, Cornwall, looking towards Land's End.

Plate 25 was obtained from Terence Soames (Cardiff) Ltd.; Plates 13, 28, 29, 32, 34 from the collection of Dr. J. K. St. Joseph, director in aerial photography, Cambridge University; the remaining 28 plates and the cover photograph from Aerofilms Ltd.

INTRODUCTION

This report is one of two separate publications which together comprise the Commission's final report on the study of coastal preservation and development in England and Wales.

The study, which was begun by the National Parks Commission in 1966, at the request of the Minister of Housing and Local Government, was based on a series of nine regional conferences with maritime local planning authorities. These conferences were held between May 1966 and March 1967, and their chief purpose, in the words of M.H.L.G. Circular 7/66, was 'to provide a firm foundation for long-term policies for safeguarding the natural beauty of the coast as a whole and promoting its enjoyment by the public'.

A report of each of the regional conferences has since been published, as well as two special study reports, *Coastal Recreation and Holidays* and *Nature Conservation at the Coast*.

The present report amplifies one of the main recommendations made in the main report—*The Planning of the Coastline*—that selected stretches of undeveloped coastline of high scenic quality should be given a special designation in order to protect their natural attractions and to encourage their use for appropriate informal recreation. The main report also deals with a range of other matters, for example, trends in the holiday and recreational use of the coast, the future of the resorts, new holiday centres, regional parks, industries, eyesores and planning policies for protecting the undeveloped coast.

* * *

The Commissioners gratefully acknowledge help received from Professor J. A. Steers in the preparation of this report.

Note

In the context of this report, *conservation* means the planning and management of resources to ensure their wise use and continuity of supply.

SUMMARY OF REPORT

In *Part 1* the need for special measures to ensure the conservation of coasts of high quality scenery is discussed. *Chapter 1* outlines the basis of the study, which relates principally to the best stretches selected from the 2,054 miles of substantially undeveloped coastline.* In *Chapter 2* mention is made of the pressures facing the undeveloped coast as a whole, and the reasons why those stretches of finest scenery are particularly vulnerable. The importance of managing these coasts in the best interests of all users is emphasised. Special measures to safeguard, for posterity, the most valuable parts they contain must also be taken. *Chapter 3* explains, with reference to examples, the need for special treatment of high quality coasts, which justify a special claim for protection if their conservation is to be secured for all time.

Part 2 describes the review of coastlines of fine scenery undertaken by the Commission. *Chapter 4* outlines the processes of survey and analysis which led to the selection of 34 areas of mainly undeveloped coast representing the best examples of coastal scenery in England and Wales. These areas extend along some 730 miles of coast and represent a little more than a third of the undeveloped coastline. In *Chapter 5* attention is drawn to the national significance of these coasts as part of the coastal heritage, and the importance of identifying them by some form of national recognition. It is emphasised that, within these areas, where the intensity of the pressure of visitors is likely to be greater than in many of the National Parks, special measures will be necessary to preserve their character. Existing legislation and designations are unlikely to prove sufficient, and the special designation of these areas as Heritage Coasts is recommended.

Part 3 of the report outlines the principles of planning and management which should be applied to coasts of high quality scenery designated as Heritage Coasts. *Chapter 6* outlines the overall planning and management policy which should be pursued, the two main objectives being to conserve the quality of the scenery and to facilitate its enjoyment by encouraging activities which rely on the natural scenery and not on man-made attractions. With proper management, the recreational use of these coasts need not be confined to only small numbers of visitors. The concept is introduced of planning for an optimum capacity: the degree of access to, and the scale of activity within, a Heritage Coast should, as far as possible, be related to the maximum number of persons and vehicles that can make use of it without damage to its beauty. Each Heritage Coast should be divided into three principal management zones.

In *Chapter 7* the need is stressed for measures to control development incongruous in an area of fine coastal scenery. A code is suggested which will enable the acceptability of non-recreational uses to be judged, and a Special Development Order is recommended to secure greater control of development than would apply outside a Heritage Coast. A further code, based on the management zones, is suggested to determine the acceptability of recreational activities. *Chapter 8* outlines the principles which should be applied to control access, by road, on foot or on horseback. The aim

---

* The basis on which the length of undeveloped coastline has been measured is explained in the Commission's publication *The Coasts of England and Wales: Measurements of Use, Protection and Development*. (See also footnote to paragraph 1.2.)

should be to avoid excessive concentrations of persons and vehicles at points which are easily damaged, or where the unspoilt or remote character of the coast or its scientific importance is threatened. Policies are suggested for the provision of car parks, scenic roads and footpaths, and for access to open country along the shore. In *Chapter 9* the provision of recreational facilities is considered, and a policy outlined which seeks to ensure that these are adequate for appropriate activities and for the enjoyment of the natural scenery, and that they are so designed and located that the character of the Heritage Coast does not suffer. Types of facility are described, and a code outlined which may be used as a test of their acceptability at specific locations.

Chapter 10 describes the measures which should be applied to secure improvement of the landscape by the removal of disfigurements and other means. The importance of co-ordinating the management policies of public agencies, and the need to consult private landowners, is emphasised. *Chapter 11* discusses the important role that information and interpretative services can play, not only in informing visitors of features of interest, but also in encouraging a variety of activities, and causing visitors to spread more evenly throughout an area. By the imaginative use of interpretative techniques, visitors may also be encouraged to take those routes which cause less damage. Greater use may also be made of existing resources without the need for the provision of new facilities.

In *Chapter 12* consideration is given to ways and means of implementing the policies outlined in the preceding chapters. There is great need for close liaison between public and private bodies and individuals to ensure that policies of management, access and development do not conflict with one another, nor with the reasons for the designation of a Heritage Coast. Much importance is attached to the role that voluntary organisations and members of the public can play in conserving the coast and supporting official policies. The need is discussed for management agreements and similar arrangements to secure the attainment of policy objectives, and recommendations are made concerning the incorporation of these policies in development plans. The establishment of a special committee of the local planning authority, with nationally appointed representatives, is recommended to implement the planning and management of a Heritage Coast. The need for a Conservation Officer for each Heritage Coast is explained. It is recommended that the salary of such an officer, together with the administrative costs of Heritage Coast management, should be eligible for special Exchequer grant.

Part One

The Need for a Policy

CHAPTER 1.
COASTAL SCENERY AS A VALUABLE RESOURCE

1.1   The coast of England and Wales offers a great variety of scenery. Its complex of rock types produces over relatively short distances a succession of rocky headlands and cliffs, sandy bays and coves, shingle beaches, estuaries and offshore islands. Much of the attraction of the coast and the esteem in which it is held by the British nation is the result of this change of character every few miles. Moreover it is this variety which makes it a unique natural resource. Despite the attractions of large stretches of inland Britain, there is something at once fascinating and unique about that part where land meets sea: there is only one coastline, and it is of limited length. (All the land within 1 mile of high water mark represents only 4% of the total surface area of England and Wales.)

1.2   Some 687 miles (or 25%)* of the coast of England and Wales are either substantially developed or earmarked for development. Of the remaining 2,054 miles of coast much of the scenery, although good, cannot be regarded as high quality, but there are many outstanding stretches. In Part 2 of this report an attempt is made to identify those areas of coast which would generally be regarded as the most attractive and beautiful. In doing so regard has been had primarily to unspoilt natural scenery, but certain man-made features such as monuments or antiquities associated with the coast have also been taken into account. Those stretches identified in this report as the most attractive are usually outside the main areas of settlement; small groups of buildings or man-made features have, however, been included where they form an integral part of the area. Major groups of buildings, such as attractive fishing villages, which may justify some measure of protection as environments in their own right, are however not considered in this report.

1.3   The references in this report to coasts of high scenic quality apply to the outstanding stretches selected from the 2,054 miles of substantially undeveloped coast. These are the areas which demand special measures to ensure their conservation. In Part 3 of this report we consider methods by which this may be achieved.

---

* Full particulars of coastal measurements are contained in *The Coasts of England and Wales: Measurements of Use, Protection and Development*. In particular it should be noted that: (1) 'Substantially developed' areas include all existing development with established or permitted use rights, together with small parcels of undeveloped or 'uncommitted' land within the developed areas, but excluding individual sites of less than 2 acres in extent. Industrial and commercial uses, as well as camping and caravan sites, are included as developed areas; (2) Where a strip of open space of at least 220 yards in depth exists between a built-up area and the shore, the frontage is not regarded as 'substantially developed'.

CHAPTER 2.
PRESSURES AND TRENDS

2.1   As has been shown in the Commission's main report* on the planning of the coastline the whole coast of England and Wales is already subject to considerable pressure. Nowhere is this pressure likely to cause damage more permanent, or a loss more complete to the nation's heritage, than on unspoilt coasts of high-quality scenery. These areas are particularly liable to damage not only from large-scale industrial or commercial developments, power installations, and the like, but also, increasingly, from the pressures of recreational use. These include the demand for holiday accommodation, caravan sites, marinas and other facilities, and also the great numbers and mobility of those who, with their cars, are able to visit the coast at any one time, whether for day trips or holidays.

2.2   Although 75% of our coastline is substantially undeveloped, the pressures on the coast as a whole imply that some of the undeveloped stretches must inevitably be made available in the future for development in one form or another. It is the purpose of this study to examine how the coasts of high quality scenery may be safeguarded.

2.3   In the past it has been argued, by those anxious to preserve at all costs what is becoming a dwindling heritage, that all undeveloped coast is worth preserving for all time against any form of intrusion or encroachment. Clearly, however, some developments for which a coastal site, large or small, is essential must be accommodated on the coast. There are also many instances where a coastal site is not essential, but merely desirable. To deny the need for a coastal site, where such a need can be proved to exist, in order to preserve undeveloped coastline simply because it is undeveloped, defeats the objectives of conservation. Many stretches of undeveloped coast have no great scenic or scientific merit and are suitable for carefully planned development, assuming a case could be made for a coastal site. Moreover, much coastal recreation can be satisfied by the concentration, at selected locations—for example, at beaches— of facilities for intensive recreational activities for large numbers, thereby easing the pressure on the neighbouring stretches. These developments, whether sited in new locations or at established resorts, are a perfectly acceptable form of development for a coastline of low or medium quality scenery.

2.4   The conservation policy we wish to promote takes current pressures into account and recognises the need to manage the coast in the best interests of all users, but at the

* The Planning of the Coastline: a report on a study of coastal preservation and development in England and Wales (H.M.S.O., 1970).

same time takes special steps to safeguard, for posterity, those most valuable (and often most vulnerable) parts. This approach accepts the need to accommodate those pressures that can be met without conflicting with the conservation of the environment, and implies, wherever possible, the full use of coastal resources, in conjunction with a strictly enforced development policy. This should exclude completely from vulnerable stretches those forms of development (including some recreational activities) which, by reason of noise, scale, traffic, or damage to the habitat, would be totally incompatible.

2.5   The increasing numbers and use of the private car, and associated problems of erosion and trampling, pose a threat to the coastal environment. It is worth remembering, however, that at present the problems which arise from direct vehicular access to the shore or cliff top by no means apply to the whole of the undeveloped coast. Considerable areas of coast are not served directly by roads or motorable tracks, and are accessible only on foot or horseback. Despite a general impression to the contrary, the visitor who prefers the motorless environment and is prepared to seek it out can still do so without much difficulty in most parts of England and Wales. Even in the height of the season secluded or unused beaches can be discovered with little effort, provided one walks to them. But without careful planning, even minor road improvements which facilitate access to those parts which cannot now easily be reached by car may introduce visitors in such numbers that the character of the area will be changed and damage to the environment may result. Those stretches of coast which are both motorless and also of high quality scenery are to be counted among the most valuable of the nation's coastal resources.

2.6   In our main report* we consider in some detail the interrelationship between planning and management policies in order to secure better conservation of our coastal resources as a whole. Past experience illustrates how piecemeal development, because of inconclusive policies, leads to changes in the environment. For example, by gradual stages, roads are widened to permit a greater flow of traffic to the coast. New roads and car parks are then constructed to meet the demand, which is in turn further stimulated and accelerated by these. Despite planning controls and protective designations more petrol filling stations, roadside cafes, caravan sites, holiday chalets, souvenir shops, poorly-designed notices and signs and other objects inevitably appear. It is not long before the natural charm and attraction of the coast is seriously impaired. With an increase in car ownership and leisure time the pace and effects of this deterioration will be even more acute in the future, and nowhere will the results be more unfortunate than along coasts of high quality scenery.

2.7   We do not propose to repeat in this report the arguments for the proposals we make in our main report to accommodate pressures for development. We envisage a general policy of concentration, revitalising the established resorts, creating new resorts, creating regional recreation areas and special areas for industrial development. These steps should absorb some of the demands on the undeveloped coast. In our main report we also stress the need to reinforce the present planning policies for the mainly undeveloped coast by clarifying them, and by more effective implementation and greater attention to details of management.

2.8   We mention also that some forms of recreational use of the undeveloped coast are compatible with the conservation of the environment provided they are sensitively planned and managed. But, as we shall demonstrate in the following chapters, a

---

* The Planning of the Coastline.

4

number of special planning and management measures will be needed in those areas of the highest quality scenery in order to cater for the human pressures to which they will be subjected. At the same time these will enable the increasing number of visitors who prefer the simpler forms of recreation (such as swimming, walking, picnicking, and nature study) to enjoy them away from the great numbers of those who prefer gregarious activities, many of whom will merely be seeking somewhere, attractive or otherwise, within easy reach of the sea.

CHAPTER 3.
THE NEED FOR SPECIAL TREATMENT

3.1　It will be generally accepted that some stretches of coastline are considerably more valuable than others as scenic resources. Many, too, are closely associated with cultural, traditional and other features not directly related to the landscape, but which nevertheless are an integral part of the coastal heritage and greatly enhance its attraction. These coasts justify a special claim for protection. In order to achieve this, it may be necessary to invoke stronger powers and measures than those required for other parts of the undeveloped coast. Further, these areas need detailed management of a comprehensive nature which will embrace landscape improvement, traffic management, control of undesirable development, and the encouragement of a number of different activities. It is important that these aspects should be considered together and not individually. There is, therefore, much to be gained by clearly identifying coasts of this kind.

3.2　To illustrate the need for special treatment it is convenient to distinguish four separate categories of coast of high quality scenery which call for different management techniques. These categories may overlap. Firstly there are those areas in early stages of defacement. Some of these would at one time have been regarded as among the most attractive in the country, but as a result of insensitive planning there is a real danger that their scenery will be permanently defaced unless urgent action is taken. These are chiefly areas of low-intensity development, often with few buildings. Attention to small points of detail, some simple form of traffic management, a scheme for landscaping, and the removal of disfigurements may achieve results without considerable expense.

3.3　Secondly, there are those unspoilt and relatively unknown stretches with limited road access. These are for the most part unspoilt because of their inaccessibility, but they are distinctly vulnerable to any road improvements which would attract more vehicles. Examples include the coast between Hartland Point and Bude and between Morvah and St. Ives (Cornwall), the north coast of Anglesey between Amlwch and Cemaes Bay, and between Porteynon and Worms Head (Gower peninsula). Some of these areas, which are among the best examples of coastal scenery to be found in England and Wales, deserve special protective measures to ensure that they may be enjoyed in their natural state. Interpretative measures* may be necessary to encourage a greater use of them by visitors on foot, on horseback or by cycle, but

* See Chapter 11.

considerable care will be needed in framing a road policy, since the wholesale improvement of vehicular access would lead to the rapid deterioration of the scenery.

3.4   The third category comprises coastal heritage features of outstanding national significance. These include extremities (such as Land's End and Lizard Point), other prominent headlands (such as Flamborough Head, Spurn Head, Start Point and Great Ormes Head) and coastal features of national renown (such as the Seven Sisters, the Needles, or the White Cliffs of Dover). Under this category we may also include cultural, historic or literary associations and other attractions, both natural and man-made. The immediate surroundings of many of our national monuments along the coast are not satisfactorily managed, and the attraction of these areas could often be enhanced by greater attention to their surroundings, better management of visitor movements, and the control of clutter, advertisements and poor quality development.

3.5   The fourth type of coast that may be distinguished in this context is an area already in some form of public or quasi-public ownership. Stretches owned by local authorities, government departments, and other similar bodies offer the greatest scope for creative management, since the ownership of the land itself is a key factor in determining permitted uses. Sometimes access is discouraged or restricted to a greater extent than is really necessary, and it is clear that, with enlightened management, many stretches of fine coastline would be capable of providing enjoyment for a larger number of visitors without destroying the environment. Some form of controlled public access will be expected and may be necessary in order to justify public ownership and support of these often extensive stretches. Indeed, significant public interest (including financial support) may be gained as a result of increased access, and this support may also make possible the securing and financing of adequate management.

3.6   In addition to these four special categories there are many other stretches of coast of high-quality scenery which justify special protection. What is needed in these areas is the concerted application of suitable interpretative techniques and much more extensive use of informative material to emphasise the rich rewards the areas offer, to enable them to be enjoyed by a wider range of visitors. This is, of course, more easily brought about in an area of coastal land which is being managed comprehensively. By promoting a greater interest in, and understanding of, these areas of coast a scarce and valuable resource will be put to better use, and at the same time the chances of winning public support for measures to protect and conserve it will be increased.

Part Two

National Review of Scenic Coastlines

CHAPTER 4.
SURVEY OF SCENIC QUALITY

4.1　In order to apply the special measures referred to in Chapter 3, it is first of all necessary to identify those parts of the coast which are of outstanding significance. Not all remote or unspoiled parts of the coast, nor all the parts of the coast included in National Parks or Areas of Outstanding Natural Beauty, nor all those parts owned by the National Trust, are necessarily of this standard; many of the finest stretches are, however, owned by the National Trust, including several added in recent years as a result of the Enterprise Neptune campaign. Furthermore, in some areas the designation of a National Park or an Area of Outstanding Natural Beauty was justified mainly by the inclusion also of attractive inland scenery.

4.2　An important criterion is that the areas chosen must be representative of the whole coast of England and Wales and must not merely reflect local or regional points of view. Many different types of coast will therefore be included. Direct comparison one with another of the parts chosen could be unhelpful. As far as possible each part chosen should not only be—in its own way—of great beauty, but also of special interest.

The undeveloped coast

4.3　The considerable volume of data assembled by the Commission from material submitted as a result of the coastal conferences* has enabled us, for the first time, to build up a comprehensive picture of land use and protection along the whole coast of England and Wales. For the coastal conferences local planning authorities submitted maps showing the extent of all land along the coast either substantially developed or committed for development. The latter category included, in addition to land allocated for development in the development plan, other land which the local planning authority had resolved to include in their future development policy.

4.4　The pattern that emerged from the conference maps shows that of the whole coastal frontage of England and Wales (2,741 miles), 2,054 miles (or 75%) is substantially undeveloped, and this is the basis of the present assessment.† We excluded those stretches already substantially developed because in these areas the predominant uses are likely to remain urban and not consistent with the conservation of fine natural

---

* See our main report, *The Planning of the Coastline.*
† For fuller details see *The Coasts of England and Wales: Measurements of Use, Protection and Development.*

10

scenery. Many portions of the undeveloped 2,054 miles consist of short stretches between towns and other settlements; for the purposes of the present study stretches less than 1 mile in length have been ignored. The pattern of undeveloped coast exceeding 1 mile in length is shown on Map 1.

Scenic merit

4.5 Those who have visited many different parts of Britain's coastline will have their own ideas of the most beautiful stretches. Some of the better-known parts are so compellingly beautiful in any weather or season that their inclusion in any national list would be readily accepted. Such a list would probably exclude many superb stretches which are inaccessible except on foot and which will have been visited only by those who have made a special effort to reach them or have had some special reason for doing so. In order to be comprehensive and include such areas the most acceptable method would be for a skilled observer to examine critically every mile of coast, irrespective of access, and record his impressions.

4.6 An assessment of this kind was carried out in 1943–1945 by Professor J. A. Steers, who defined three categories of coastal scenery: exceptional, very good, and good. The first two categories form the basis of the present assessment. The boundaries of the areas in question have been subsequently revised and amended by Professor Steers and now take account of major changes in the pattern of developed coastal land which have taken place since the original survey. The areas have also been inspected since the coastal conferences. The boundaries have been further adjusted to allow for (i) any intrusions of a minor nature which cannot easily be removed,* and (ii) any features of special significance near the extremities of the area in question, which justify inclusion in a defined stretch of fine scenery.

Assessment

4.7 Those stretches of high quality scenery which coincide with the undeveloped stretches exceeding 1 mile in length (shown on Map 1) form the basis of our study. The 34 main areas (shown on Map 2) have been further adjusted to include certain small settlements in order to secure comprehensive planning. Descriptions of the main features of these areas are given in Appendix 1.

4.8 In total the 34 areas extend along some 730 miles of the coastal frontage: this represents 26·6% of the whole coast of England and Wales. A little more than 19 miles (2·7% of the 730) are substantially developed; this amount represents the settlements which are included within the boundaries for the reasons already mentioned.† The remaining 97·3%, which comprises the undeveloped parts of the 34 areas, represents 34·6% of the total undeveloped coast of England and Wales.

4.9 We regard the stretches we have shown on Map 2 as the best examples of coastal scenery in England and Wales. They include examples of most types of coastal environment.

* Although not recorded on the local planning authorities' maps as 'substantial development' many of these would have been included as eyesores or disfigurements.

† The substantially developed portion represents 2·8% of the total developed coast of England and Wales.

5.1 The 34 areas selected in the way described in Chapter 4 form the basis of our recommendations for a special category of coast of high quality scenery. From a national point of view, these stretches deserve special measures to conserve their landscape quality and resist the pressures which threaten to destroy their character. (In the following chapters we show that such measures are not inconsistent with the use of these areas for simple forms of recreation.) We consider therefore that these coasts should be identified by some form of national recognition.

5.2 These areas must be accorded special treatment. The existing planning and countryside legislation does not ensure either adequate protection or positive management. It is essential that legal and executive decisions be taken at national level to guarantee that sufficient strength of purpose exists at local level to implement both protective policies and detailed management.

5.3 Existing designations, such as National Parks and Areas of Outstanding Natural Beauty, were introduced for the purpose of conserving high quality landscape. By reason, however, of size, character, and the nature of recreational demands, areas covered by these designations are quite different from the coastal environments considered in the present context. In the latter the detailed management of relatively small areas under intensive pressure is the key factor. In the coastal part of a National Park or of an Area of Outstanding Natural Beauty, the pressures on the narrow coastal strip are quite unlike those which apply inland, and the need for careful management is much greater. Moreover there are extensive stretches of coastline within National Parks and Areas of Outstanding Natural Beauty which have been designated because of their association with adjoining inland areas of high quality scenery but which would not have been selected solely on the basis of their coastal qualities. Parts of existing designated areas do not reach the standards we have set, and there are also superb coastlines outside National Parks* and Areas of Outstanding Natural Beauty. We are convinced, therefore, that our selected coasts must be distinguished separately.

5.4 About 9% of the land surface of England and Wales is covered by National Parks and a little over $7\frac{1}{2}$% by designated Areas of Outstanding Natural Beauty. The area of the smallest National Park is 225 square miles, while the average size of an Area of Outstanding Natural Beauty is of the order of 150 square miles. If our

* Of the 34 areas selected, six fall within National Parks.

proposals are accepted little more than 1 % of the land surface will be covered by the 34 areas we have selected.* The justification for strict measures to implement traffic management and similar schemes would only apply to a relatively small proportion of this total.

5.5   Although we regret the need for another designation we consider it essential that the areas selected in this report should be designated, nationally, as Heritage Coasts.† We have chosen this title from a variety of alternatives (including Coastal National Parks, Coastal Areas of Outstanding Natural Beauty, National Seashores and Coastal Country Parks) because we believe it expresses, better than any other, the objective behind the designation. This is to ensure that, for this and future generations, the best of our heritage of coastline is protected from exploitation and the changes that follow development. We must also stress that we believe that the policies of conservation practised within these areas will facilitate and enhance their enjoyment by the growing numbers who value good scenery and enjoy simple forms of recreation.

5.6   We consider that impending changes in the form and structure of development plans, and the increased degree of public participation in planning matters envisaged in the Town and Country Planning Act 1968, afford excellent opportunities to support the implementation of our proposals for these coasts. Before considering the suggested methods by which this may be achieved we shall examine in some detail the principles of management we envisage.

---

* This assumes a uniform width of 1 mile above high water mark.

† Throughout the remainder of this study these high quality coasts, of which the 34 areas represent the Commission's selection, are referred to as Heritage Coasts. Although detailed local considerations are bound to affect the eventual choice of boundaries, and may justify the inclusion of additional stretches not referred to in this study, it is hoped that the greater part of each of the 34 areas selected will be included in any designation.

Part Three

Heritage Coasts

CHAPTER 6.
OVERALL POLICY AND PRINCIPLES OF MANAGEMENT

6.1 The principal qualities possessed by the proposed Heritage Coasts are their unspoilt character, their exceptionally fine scenery and their heritage features. The two main planning objectives for these coasts should be (1) to conserve in its natural state as far as possible the quality of the coastal scenery: this implies careful management and, where applicable, landscape improvement, and (2) to facilitate and enhance their enjoyment by the public by the promotion and encouragement of recreational activities consistent with the conservation of fine natural scenery.

6.2 In seeking to conserve the environment the aim should be to make the wisest use of all coastal resources rather than to preserve scenic stretches for their own sake or to discourage access thereto. Care will also be needed to ensure that the uses to which these coasts may be put in the future do not alter their character so completely that the quality of their natural scenery is impaired.

6.3 In seeking to encourage and promote simple forms of recreation the aim should be to facilitate the enjoyment of the natural qualities of the coast without introducing man-made attractions such as amusements and entertainments. No attempt should be made to provide large-scale or sophisticated recreation facilities of the kind normally found in a resort. There will, however, be parts of Heritage Coasts which already attract large numbers of visitors because of good beaches, or fine headlands, or for some other particular reason—for example, Land's End, Beachy Head, and the coloured sands of Alum Bay. These must not be regarded as resorts or settlements or even as substantially built-up areas. They may, with careful management, accommodate even greater numbers than they do at present without any serious detriment to the general character of the Heritage Coast. But any attractions must be based on the natural qualities of the area and no attempts should be made to introduce artificial ones. Given effective management, therefore, recreational use of a scenic coast need not be limited to a small number of visitors. The importance of management as a tool in conserving the environment is well illustrated in this way.

6.4 To ensure the effective management of each area of Heritage Coast a number of fundamental principles and policies must be applied. Seven main principles are suggested to achieve this.

(1) *Determination of intensity of use.* The policy of management and the scale on which facilities are provided throughout the Heritage Coast should be related directly to an acceptable level of use. This must reflect the maximum number of persons, vehicles and recreational activities that the coast is able to withstand without serious damage to the environment. The overall figure for any length of coast will be deter-

16

mined by the carrying capacities of its various parts, which may vary according to their ecological stability and landscape qualities. This concept of optimum capacity should govern the pattern of access, and the scale and provision of facilities such as the number and size of car parks. Positive measures must be applied to discourage and curtail any demand which would lead to the acceptable level being exceeded and the resources being over-used. (This is considered further in paragraph 6.5.)

(2) *Determination of management zones based on different intensities of use.* The acceptable levels of use for each part of a Heritage Coast should be determined. There should be distinguished, first, those parts of the area more appropriate for relatively intensive use (such as popular beaches and major centres of attraction), second, those stretches more suitable for low-intensity use, and third, all other parts which fall between these two categories. This implies the division of a Heritage Coast into 'intensive', 'remote' and 'transitional' zones. Existing settlements should be regarded as a separate category. (This is considered further in paragraph 6.16.)

(3) *Control of development.* Rigorous control should be exercised over all forms of development which are either incongruous in an area of fine natural coastal scenery by reason of scale, siting, design, noise, disturbance, and traffic, or which adversely affect heritage features and unspoilt or remote stretches of coast. (This is considered further in Chapter 7.)

(4) *Regulation of access.* Access, for pedestrians and vehicles, should be regulated in such a manner as will avoid excessive concentrations of persons and vehicles at those points which are most easily damaged, or where the unspoilt or remote character of the coast or its scientific interest and importance is threatened. (This is considered further in Chapter 8.)

(5) *Landscape improvements.* Schemes should be initiated to improve and enhance the appearance of the landscape by means of restoration, landscaping, tree planting and the removal of disfigurements. (This is considered further in Chapter 10.)

(6) *Diversification of activities.* Attempts should be made to emphasise the opportunities for recreation in the area, especially those which permit a greater use to be made of the existing resources (such as antiquities, viewpoints and other features of interest) without the need for any substantial increase in man-made facilities, and also those which do not involve such close links with the motor car. As a result the concentration of visitors at relatively few points (mainly near car parks) will be diluted and will be spread more evenly over the whole Heritage Coast, thus avoiding the need for opening up to motorists parts which cannot at present be reached by car. Recent experience suggests that, when suitably encouraged to do so, more and more visitors are now willing to leave their cars and walk to the beach, or sightsee, or follow guided walks and nature trails. The key to this encouragement lies in good information services and in the enlightened application of interpretative techniques, designed not merely to persuade the visitor to leave his car but also to demonstrate that it will be well worth his while to do so.

(7) *Provision of interpretative services.* In addition to encouraging the diversification referred to in (6) above, steps should also be taken to promote, by means of information services and a wide variety of interpretative techniques, a closer understanding of, and interest in, the coastal environment, and in particular its natural features, flora and fauna, and any features that may be of architectural, archaeological and historic interest. (This is considered further in Chapter 11.)

Determination of intensity of use

6.5 The first principle outlined in paragraph 6.4 implies the ascertainment of an optimum capacity above which the pressures on the environment will be so great as to cause serious damage. This approach is essential to ensure that those who come to enjoy a particular stretch of coast may not do so in such numbers as will destroy the very qualities they come to seek. The acceptable level of use and the right types of recreation are difficult to define. If the environment is not to be steadily eroded, clearly a level of saturation must be determined for each distinctive resource within a Heritage Coast. Although the rate of deterioration can usually be checked, it may sometimes be difficult to determine the point beyond which restorative measures are either ineffective or prohibitive on grounds of expense. Because of the likely growth of the population, and the increase in car ownership, it will become increasingly difficult to determine the scale of facilities to be provided unless this concept is applied.

6.6 The carrying capacity of a resource may be considered in two distinct ways. Firstly there is the ecological effect on the environment arising from the numbers of people and their activities. Above a certain level, the habitat will suffer as a result of trampling, disturbance to wild life, pollution, erosion of sand dunes, and so on. Secondly there is the visual or environmental effect of people on the landscape; too many visitors will, as a result of excessive noise, overcrowding, and car-parking, destroy the character of the area they come to enjoy.

6.7 A number of points need emphasising. In the first place the carrying capacity of a particular resource cannot be absolute. The presence of even a small number of people will have a certain effect on the environment, but ten times that number may result in conditions that are intolerable. The optimum number will be somewhere between these two extremes. Moreover the effects on different kinds of resources will vary. In some habitats appreciable damage may be caused by the presence of a large number of people over a single weekend, but in other areas it may take a whole year to produce serious damage. The effect of people on the environment will also depend on other factors such as the nature of their recreational activities or whether they are accompanied by cars or dogs.

6.8 The acceptable capacity for any particular habitat will depend on which of the two main effects, either ecological or visual, is the more important. An area of sand dunes may, for instance, absorb fifty persons without visual damage, but twenty may be the maximum acceptable if erosion is to be prevented. A further difficulty arises in measuring the visual effects which, being subjective, will arouse different responses among individuals. It may be possible, by more intensive or elaborate management, to increase the actual capacity of a particular resource, for example by planting trees to reduce the visual impact of a large number of visitors or cars, or by providing surfaced paths on a cliff subject to erosion and confining access to these.

6.9 The determination of what is an acceptable capacity for any particular resource will thus depend on three factors, the tolerable degree of ecological disturbance, the desired level of solitude or of acceptable visual intrusion, and the available funds for investment in management. At the current stage of research little progress has been made in establishing methodology for assessing the capacity of resources in terms of either ecological or visual effects. More research is needed, because it is fundamental to the conservation of an environment under pressure to be able to examine these effects and suggest solutions. This is a task to which we can contribute under our research powers.

18

6.10   In the meantime it is best to make a cautious approach and err on the side of under-estimating the capacity of a resource. This can be done by considering the following factors:

  (a) the extent of present use of the resource;

  (b) the potential use of the resource;

  (c) the extent to which the present use of the resource is already showing signs of deterioration;

  (d) the nature of the landscape and its ability to absorb both vehicles and people —for example, its scale and characteristics, whether wooded or open, or flat or undulating;

  (e) the vulnerability of the resource to erosion and the likelihood of damage to its scientific interest;

  (f) the relative importance of the scientific interest of the resource, and the degree of loss sustained if its scientific value was spoiled;

  (g) the degree to which the resource is already accessible to motor vehicles;

  (h) the use of the resource in relation to adjacent car parks, lavatories, cafes, and so on.

6.11   However, until capacity can be determined quantitatively a flexible approach must be adopted. There should be a readiness to adjust policy experimentally as a result of the carefully monitored observation of (1) the behaviour of visitors and the way they respond, and (2) the effects of different levels of use on the environment.

6.12   The process outlined in paragraphs 6.10 and 6.11 will help to determine the optimum figures for each resource, and, as a result, for the Heritage Coast as a whole. We show later in this report how these figures should determine planning and management policy for the Heritage Coast.

6.13   When desired optimum figures for the Heritage Coast and its component resources have been determined, various measures may be necessary to control the intensity of use. These will include:

  (a) limiting the width and capacity of roads so that the number of vehicles driven into the area will not cause damage (close liaison must be maintained with local highway authorities regarding proposals for road improvements);

  (b) providing a fixed number of car parking spaces and preventing parking at other places by means of banks, ditches or posts;

  (c) charging discriminatory parking fees to discourage long-term parking or parking close to beaches;

  (d) controlling vehicular access to certain areas by permit for those who either live or work in the area or have a special need to use motor vehicles, or are members of recognised clubs and societies of a specialist nature (the number of permits that could be issued over a given period would be limited);

  (e) siting car parks away from the more vulnerable localities (for example a short distance inland from a beach) and providing access instead by footpaths or tracks;

  (f) limiting the capacity of facilities such as cafes and restaurants;

  (g) providing efficient information services so that persons wishing to enter the area may have up-to-the-minute information on the availability of parking spaces and other facilities (thereby directing visitors to less used parts and discouraging them from going to areas where there is no room for them).

6.14 Although the peak demand lasts for a relatively short period in the summer and these measures may well have little practical effect at other times, it is, nevertheless, almost certain that the principle of admitting only a maximum number of people to an area will be disliked. The consequences of trying to accommodate too many cars and people are however even more unpleasant. If any parts of the coast of England and Wales are to remain unspoilt, there is no real alternative. Not only is a unique natural resource at risk, but never before have the pressures which threaten to destroy it been so great, nor so likely to increase. The argument is accepted without question in hotels, theatres, restaurants, art galleries and exhibitions and those who wish to visit such places at peak periods accept the need to make early application or arrive in good time to be sure of securing a place. If the beneficial effects of such a policy when applied to a piece of coast can be adequately demonstrated so that people understand the reason for limiting numbers, initial opposition may give way. The reaction of residents and visitors to the controlled parking experiment in Polperro, introduced in 1968, has shown that this is possible.

6.15 The basic principle of management throughout all Heritage Coasts should therefore be to plan for an optimum level of use. This principle could also be applied elsewhere, and would have the effect of channelling recreational demands to those parts of the coast which are able to absorb them more readily. In this way it would play an important part in the conservation of the whole coastline.

Determination of management zones based on different intensities of use

6.16 The second principle outlined in paragraph 6.4 implies the need to adopt, for the purposes of management, some rudimentary form of zoning within those parts of a Heritage Coast *outside existing settlements*.* The zones suggested are (i) *intensive zones*, consisting chiefly of the more popular beaches and other major centres of attraction of the type referred to in paragraph 6.3 above, and catering for the gregarious activities of a relatively large number of people (but only those activities which will have a minimum effect on the landscape), (ii) *remote zones*, comprising the less accessible stretches referred to in paragraph 3.3 above, and catering for those activities likely to be undertaken in small numbers by individuals or groups, and (iii) *transitional zones*, serving principally as a buffer zone between the other two. Not all Heritage Coasts need include all three types of zone.

6.17 The management policy for an *intensive zone* should ensure that the facilities provided are sufficient in number and are sited and designed in such a way that they have a minimum effect on the beauty of the Heritage Coast but enable the maximum public enjoyment of the natural qualities of the area to be obtained. These zones may be regarded as intensive on account of the numbers of persons they accommodate rather than because of the scale of activities or the nature of ancillary facilities required. Except during the peak holiday season they may well be indistinguishable from other parts of the Heritage Coast.

6.18 Intensive zones might include beaches and other undeveloped stretches of coast which are already popular, and also those which are specially suitable for intensive use but have not yet become well known. In both cases they will not have attracted the clutter of supporting facilities which characterises a resort. These zones would cater for intensive recreation of a simple nature, such as swimming, sunbathing, walking, sightseeing and picnicking. Activities which detract from these forms of

* See also para. 6.24.

20

recreation because of noise, danger or other disturbance would be excluded. Facilities would not normally be provided for overnight accommodation. Apart from the car parks and basic services such as lavatories and simple catering facilities, which would be carefully sited, the activities would rely almost entirely on the natural resources of the area—i.e. the beach or headland concerned.* Those who require more sophisticated facilities plus the full range of piers, amusements and entertainments would need to look instead to the resorts.

6.19   A *remote zone* would aim to retain a selected stretch of relatively inaccessible and unspoilt coast free from vehicles. The justification here is twofold: firstly to enable the protection of fragile habitats which are very liable to damage both by vehicles and large numbers of people, and secondly to provide for those whose enjoyment depends to a large extent on relative solitude and the absence of vehicles. There is no justification for allowing cars to reach every mile of coast in England and Wales, and to spoil the pleasures of an increasing number who demand peace and quiet. Large sections of coast are, in fact, under-used by motorists even where there is good vehicular access.

6.20   Remote zones would include areas so little touched by man that a feeling of solitude is apparent. They would cater for individuals or small groups who seek simple pursuits not requiring service facilities or road access (e.g. walking, ornithology, cliff climbing, field studies, etc.) as well as those who wish to picnic, sightsee or sunbathe in secluded spots. These zones would also meet the needs of the increasing number of motorists who seek recreation away from the car. The rise in the popularity of self-guided nature trails† in recent years is evidence of this trend. The more extensive sandy beaches, some of which are at present relatively inaccessible, may be more appropriate in an intensive zone, but a remote zone could well include many smaller coves and inlets where those who prefer seclusion may find it, with a small degree of effort. To maintain the intensity of use at a low level some remote zones would need to be fairly extensive, perhaps 3 miles or so in length. (For examples of possible remote zones see paragraph 3.3.)

6.21   The remainder of the Heritage Coast outside existing settlements should form a predominantly undeveloped *transitional zone*. Management policy should aim to group recreational and service facilities in selected places which will cause the least damage to the environment, and should retain, allowing a reasonable degree of road access, the stretches in between as unspoilt, natural scenery. In many cases this zone would form the greater part of a Heritage Coast, and would cater for those such as motorists who may wish to picnic, stroll and enjoy the scenery in quiet surroundings but who would not be prepared to walk any great distance from the car, and also for those who wish to sit in or near their cars and enjoy the view from a cliff edge or headland rather than spend a day on a crowded beach.

6.22   The first stage in the process of formulating an overall plan for a Heritage Coast should therefore be to divide the coast into management zones based on a study of the capacity of existing resources and their characteristics. Appendix 2 to this report suggests a method of approach.

* Examples of intensive zones might include Beachy Head, The Lizard, and the coast between Bamburgh and Seahouses (Northumberland).

† i.e., where the visitor is led along a defined route on foot by means of an explanatory pamphlet and a series of numbered posts or markers on the ground.

6.23 The determination of an inland boundary of a Heritage Coast will depend largely on the nature of the topography and the pattern of land use. It is difficult, therefore, to generalise and suggest how this may be done, because detailed local considerations and the provisions of the development plan will dictate what is a realistic limit. It is envisaged, however, that a Heritage Coast would be a fairly narrow coastal strip, usually not more than perhaps $\frac{1}{2}$ to $1\frac{1}{2}$ miles wide.

6.24 The need to include some small settlements within the boundary of a Heritage Coast has been mentioned in paragraph 4.7 and all that has been said so far in this chapter applies to substantially undeveloped land. For purpose of comprehensive management it is clear that policies covering such matters as access and landscape conservation should apply equally to the settlements within a Heritage Coast, although it may not be desirable for these settlements to be included in any of the three zones suggested. Individual local planning authorities have, however, widely differing rural settlement policies and, as in the case of the Heritage Coast boundary, the particular local considerations which apply should also determine precisely where the boundary between a settlement and an adjacent zone should lie, and whether for instance a village 'envelope' should contain merely the existing buildings or allow for future growth.

6.25 Having decided broadly the desired levels of use and the zones in which these apply, it remains to consider how the management policies should operate in each zone, in relation to access, control of development and the provision of recreational facilities. These aspects are considered in the chapters which follow.

CHAPTER 7.
LAND USE AND DEVELOPMENT

7.1   Whatever form the management of a Heritage Coast may take, it is unlikely to be effective unless allied closely to a firm policy of land use and development control within the statutory planning framework. That policy should conserve the special quality of the landscape, safeguard features of scientific interest and support agriculture, forestry, and simple forms of recreation in such a way that they do not conflict with one another. In remote zones there will be strong opposition to any new development, but in the intensive zones it may be necessary not only to accept but also to promote new facilities.

7.2   The character and significance of Heritage Coasts demand some special measures to reinforce this policy. Unfortunately the protective policies of local authorities vary widely in effectiveness from area to area. Sometimes the policy which covers large inland tracts of countryside under minimal pressure applies also to the narrow coastal strip, where the pressures and possible damage arising from them are greatest. Within the coastal strip, incongruous development should be more firmly resisted, and categories of acceptable uses and development should be clearly defined.

Non-recreational land uses

7.3   As far as non-recreational land uses are concerned, those which are considered acceptable in a Heritage Coast include:

(i) uses already established, those which are the subject of outstanding planning consents, and those which, by reason of their scale, their contribution to the local economy or their impact upon the landscape would not justify the compensation involved in a revocation or discontinuance order;*

(ii) agricultural uses;

(iii) forestry uses;

(iv) the use of land for nature conservation or field studies;

(v) development for the purposes of coastal navigation or safety (e.g. life-boat service, lighthouses, coastguards, sea rescue).

* This assumes that the existence of large non-conforming uses would be taken account of in drawing up the Heritage Coast boundary.

7.4 The following criteria should be applied to any new buildings, works or other forms of development required in connection with the above uses:

(a) the siting of all buildings and structures should receive more than the usual care to ensure they do not spoil the landscape;

(b) where possible all buildings and similar works should be grouped at established centres or close to existing buildings and works;

(c) special care should be taken in scale and design to ensure that all buildings harmonise with the landscape and enhance, and not detract from, the character of the Heritage Coast;

(d) wherever possible local materials should be used;

(e) particular care should be given to details such as the treatment of open spaces around buildings, means of access, and the treatment of walls, fences or other forms of enclosure.

7.5 Any non-recreational uses other than those listed in paragraph 7.3 should be regarded as unacceptable, and permission for development* should be refused unless they are of such national importance as to justify not only the need for a coastal site rather than one inland, but also the need for a site within a Heritage Coast rather than a coastal site elsewhere. In considering whether the national loss sustained by using an alternative site is greater than that which would arise from the spoliation of a Heritage Coast, the following factors should be taken into account:

(i) the likelihood of noise and disturbance, and their effects on the character of unspoilt and secluded stretches of coast and on areas and features of scientific interest;

(ii) the likely effect of increased traffic, its effect on approach roads, and the effect of car parks;

(iii) the effect the proposed development, and the number of persons occupying or visiting the site, would have on features of special interest within the Heritage Coast;

(iv) the effect on the accessibility of the coast, and especially the extent to which access routes and parking places provided for the site may or could be used by visitors seeking recreation;

(v) the defined management zone within which the site falls: account should be taken of the appropriate policies for that particular zone (in remote zones, for instance, proportionally greater attention will need to be paid to any development which alters the pattern of access, bearing in mind the need to retain the special character of these very limited areas).

7.6 Problems sometimes arise as a result of operations which do not constitute development requiring planning permission. Agricultural and forestry operations, for instance, are expressly excluded from planning control by Section 12(2) of the Town and Country Planning Act 1962, as also are road improvements carried out within the boundaries of a road by a local highway authority. Other forms of development set out in Schedule 1 to the Town and Country Planning General Development Order 1963 are permitted by the Order and do not require express planning consent.

* 'Development' is defined by Sec. 12(1) of the Town and Country Planning Act 1962 as the carrying out of building, engineering, mining or other operations, or the making of any material change of use.

Examples of operations which may conflict with the conservation of a nat
scape include the erection of agricultural buildings and works (Class V
buildings and works (Class VII), road improvements carried out adj
by a local highway authority (Class XIV), temporary uses such as ca:
28 days (Class IV(2)), recreational uses by approved organisations (C.
development by statutory undertakers such as harbour boards, water boards .
electricity and lighthouse undertakings (Class XVII).

7.7   Operations such as these could easily spoil a Heritage Coast. In the absence of
special provisions a local planning authority is powerless to prevent them, nor may it
advise on such important aspects as design and siting, nor may it comment on proposals which may conflict directly with an overall management plan (for example
road improvements which facilitate a greater flow of traffic or involve the removal of
stone walls or hedgerows). Some local planning authorities meet this problem by
working closely with local developers, landowners, agriculturalists and other interested
parties by means of a countryside advisory committee, and rely on voluntary consultation and goodwill to achieve the necessary co-operation. There is much to be said
for this approach. Similar arrangements operate successfully in some of the National
Parks so far as forestry operations are concerned. By this means a local planning
authority may be able successfully to resolve differences of opinion and achieve a
degree of co-operation which the enforcement of stronger measures (which a local
planning authority would in any case be reluctant to invoke) would be unlikely to
achieve. A similar degree of close co-operation is necessary with local highway
authorities to ensure that road improvements are examined in the light of the overall
management plan for the Heritage Coast. However, insofar as concerns the control
of development (though not the management of land), we doubt the lasting effects
of a system of voluntary co-operation if the basic policy of safeguarding these areas
for all time is to be implemented.

7.8   We consider therefore that stronger measures should apply in Heritage Coasts to
ensure that the management policy is not frustrated if voluntary liaison should prove
ineffective. This applies especially to agricultural buildings, which are exempt from
the normally strict control standards applied even to minor buildings in areas of high
landscape value.

7.9   Some additional powers of control within certain areas of high landscape value
are already available to local planning authorities. The Town and Country Planning
(Landscape Areas Special Development) Order 1950, for example, provides that within
relatively small defined portions of three of the National Parks no agricultural or
forestry development which would normally be permitted by the General Development Order may be carried out without notification to the local planning authority,
who may make requirements about the design and external appearance of the buildings proposed. The question of siting, however, which may well be of the greatest
importance from the landscape point of view, does not come within the scope of this
arrangement—a serious limitation.

7.10   Control over permitted development may be more effectively strengthened by
making a direction under article 4 of the Town and Country Planning General
Development Order 1963, the effect of which is to enable the local planning authority
to exercise control by granting or refusing planning permission for development of
a specified class which would otherwise be permitted under the Order. The experience
of a number of local planning authorities has shown that the Minister is generally

c

unwilling to approve directions covering wide areas of coastal land, and the majority of those approved apply to particular sites where the lack of any control over a specified class of permitted development has already become such a serious problem as to warrant a direction being made. This policy frequently discourages local planning authorities from submitting directions for Ministerial approval and often precludes the making of a direction in cases of the greatest importance, namely where the pressures are such that considerable damage to the landscape is imminent but has not yet occurred.

7.11   In respect of agricultural buildings, Circular 39/67 sets out the criteria the Minister will take into account when considering whether approval should be given to article 4 directions covering more extensive stretches in scenic areas, of which unspoilt stretches of coast are cited as an example. A local planning authority has to establish that the area in question is particularly vulnerable to damage by the indiscriminate siting of farm buildings on account of the topography or the special quality of the scenery: factors which apply to every stretch of Heritage Coast. Having regard, therefore, to the limited extent of coastal land which would be covered by Heritage Coasts, and the outstanding quality of their scenery, we conclude that if such an extensive article 4 direction were requested by a local planning authority the Minister would not be justified in withholding his approval.

7.12   We recommend therefore that provision should be made for a Special Development Order applying specifically to Heritage Coasts, having the same effect as an article 4 direction and covering a wide variety of the classes of permitted development set out in Schedule 1 to the Town and Country Planning General Development Order. We do not propose that such a Special Development Order should necessarily apply to villages and small towns within the Heritage Coast.

7.13   Another measure already available to local planning authorities is the definition of an area of high-quality environment as an area of special control of advertisements (by regulations under s.34 of the Town and Country Planning Act 1962). In such an area the express consent of the local planning authority is required for a wider range of advertisements than is otherwise the case and the types of advertisement which may be permitted are limited and are clearly set out. Heritage Coasts are easily spoiled by ill-sited and insensitively designed advertisements and signs. There is therefore much to be said for the application of the measures described in all Heritage Coasts, each of which should constitute an area of special control. Consequently we recommend that the Special Development Order (see paragraph 7.12) should provide for the definition of each Heritage Coast as an area of special control.

Recreational land uses

7.14   As far as recreational land uses are concerned there are two aspects to consider, first the activity itself, and second the provision of facilities to support it. Obviously unless the activity is acceptable within a Heritage Coast permission should not be granted for the supporting facilities such as car parks, club huts, changing accommodation, and equipment stores. There should be a different policy in each zone of a Heritage Coast in order to maintain the characteristics of these zones. (The provision of recreational facilities is considered in more detail in Chapter 9.)

7.15   The policy governing recreational development should be threefold: first to encourage those activities which rely wholly or mainly on the natural resources of the coast and have little or no impact on the landscape, second to confine those activities which are likely to conflict with each other to specified parts of the coast to reduce or eliminate this conflict, and third to discourage and resist the introduction of activities which, no matter how desirable in themselves, are likely to damage the character or scenery of the Heritage Coast.*

7.16   The determination of what are acceptable activities should be based on the nature of the management zones referred to in paragraph 6.16. The model Chart of Acceptable Activities (see page 28) illustrates one method of achieving this. Ideally a more sophisticated approach may be desirable which takes account of the special characteristics of each individual resource within the Heritage Coast. The demarcation of such resource zones, which would also apply to non-recreational uses, will take time and will involve detailed consideration of the characteristics, vulnerability, potential uses and evaluation of each resource. This will call for a degree of specialist knowledge of agriculturalists, foresters and ecologists and may well best be undertaken by an inter-disciplinary team composed of representatives from the local planning authority and the specialist interests mentioned. The joint exercise carried out in 1967 within the East Hampshire Area of Outstanding Natural Beauty† illustrates one approach.

7.17   In the meantime the adoption by local planning authorities of a simple chart similar to the one illustrated (amended to apply to the Heritage Coast in question), and forming part of the development plan, would help them to determine the acceptability of recreational uses clearly, and would also help to safeguard those areas that need to be protected from the more intrusive activities. Where control of an activity inappropriate in a particular zone is ineffective, other measures, such as the introduction of byelaws or speed limits, and a system of wardening, may be necessary. This may apply in certain zones to activities such as power boating, water ski-ing, motor sports, the flying of model aircraft and the playing of portable radios.‡ The Chart of Acceptable Activities agreed for any particular stretch of Heritage Coast will also indicate the range of activities to be encouraged and promoted as part of the management policy.

7.18   It is suggested therefore that local planning authorities should adopt and incorporate into their development plans measures such as these to determine which recreational activities are appropriate in each zone. For the reasons mentioned in paragraph 6.24 many settlements within a Heritage Coast are excluded and will need special consideration.

* i.e., by reason of scale, noise, traffic or disturbance to wild life arising from the activity itself or from the ancillary facilities that would be required, or the numbers taking part.

† Described in the report *East Hampshire Area of Outstanding Natural Beauty: a study in Countryside Conservation.*

‡ The extent to which it is legally possible to control activities on the beach, foreshore and open sea is considered in paragraph 8.31.

CHART OF ACCEPTABLE ACTIVITIES

...ty	Remote Zones	Transitional Zones	Intensive Zones
...games	B	A	A
...g (motorised)	D	D	C
Camping (lightweight)	C	B	B
Canoeing	C	A	B
Cycling	C	A	B
Field Studies, nature study and Outdoor Education	B	A	B
Flying model aircraft	D	B	C
Golf	D	B	B
Motor sports	D	D	D
Picnicking	B	A	A
Playing portable radios	D	B	B
Pleasure Motoring	D	B	B
Pony Trekking	C	A	B
Potholing	B	A	B
Power Boating	D	C	B
Riding horses	C	A	B
Rock and Cliff Climbing	B	A	B
Rowing	B	A	A
Sea Angling	A	A	A
Swimming	A	A	A
Sailing	B	A	A
Sand Yachting	D	D	D
Skin Diving	C	A	B
Sunbathing	A	A	A
Surfing	B	B	A
Walking	A	A	B
Water Ski-ing	D	D	D
Wildfowling	C	B	D

Codes:

A   indicates the zone or zones within which the activity is considered most appropriate and in which its promotion or development should be encouraged.

B   indicates a zone within which the activity itself is acceptable, but for various reasons (for example the need for car parks, buildings or structures which would themselves be inappropriate in the zone concerned, or because of the limitations of the zone itself) is not the ideal location and no steps need be taken to promote the activity.

C   indicates a zone within which the activity, subject to certain safeguards, may be appropriate on a limited scale. Acceptability will depend on the nature of the locality and the scale of the activity envisaged. Steps need not be taken to promote the activity.

D   indicates a zone within which the activity is wholly inappropriate and should be resisted. This category includes activities characterised by excessive noise or disturbance, or those which require large tracts of land or water for satisfactory performance or maximum enjoyment.

CHAPTER 8.
ACCESS

8.1 The degree of access within a Heritage Coast afforded by the roads, parking places, footpaths, bridleways and open country is a key factor in determining the levels of its use. The management policy outlined in Chapter 6, and the concept of optimum capacity, are likely to succeed only as a result of careful management of vehicles, pedestrians, horse riders, cyclists and other users in accordance with an agreed access policy. It is also essential to ensure that the rights of access of local residents within a Heritage Coast are not infringed.

8.2 The main elements which facilitate access within a Heritage Coast are:

(i) *public roads* on which vehicular traffic is either passing through the Heritage Coast or seeking access to parts of it. (For access policy, including also elements (ii), (iii) and (iv) below, see paragraph 8.3 of this Chapter.)

(ii) *private roads* serving isolated communities, farms, lighthouses, and private estates. These could be used by any motorist in the absence of locked gates, deterrent notices or other measures of control. (Traffic on public and private roads may disturb wild life, or damage the habitat on account of noise and fumes. Too much traffic may also be an eyesore.)

(iii) *tracks used by vehicles* which lead to beaches or open country, and which, although not public rights of way for vehicles, could be used by the more adventurous motorists. The effects of this traffic will be similar to that on the roads. There may also be serious erosion of the surface and problems may arise from the obstruction of narrow tracks by vehicles. Where it is desirable to confine access along these tracks to service and rescue vehicles, and vehicles of local residents, special measures will be necessary to deter or restrict other vehicles.

(iv) *parking places* for vehicles—lay-bys, properly constructed car parks, and fields or other areas in temporary use for parking during peak periods. Problems of noise, fumes, erosion and visual intrusion may arise, but more significant than these will be the numbers of persons deposited within the area at these points. Since most visitors to a Heritage Coast are likely to travel by road, the siting and capacity of parking places is of the greatest importance in determining the level of use and activity in each management zone.

(v) *public footpaths*, which usually cross agricultural land in private ownership, and offer access to beaches and cliffs. Paths may be made or diverted more easily than roads, to lead walkers away from fragile habitats or to direct them towards specific areas or features, such as viewpoints. Conflict between walkers and agricultural interests may however arise. A suggested policy regarding footpaths in a Heritage Coast is outlined in paragraph 8.16.

(vi) *bridleways* for horse riders, walkers and, subject to local byelaws, by cyclists. Factors similar to those mentioned above in connection with footpaths will apply. Horses' hooves and cycle tyres may also cause erosion, and may make walking uncomfortable. The appropriate policy for bridleways is considered in paragraph 8.23.

(vii) *beaches and open country*, (including cliff tops, headlands, sand dunes, downland, heath, commons and other uncultivated land where the public are allowed to wander freely and are not confined to specified paths and tracks). Within a Heritage Coast, these, together with the sea itself, will be the main resources for recreation and are the places to which roads, tracks and paths should lead. The character of the coast will depend very largely on the extent to which public access to these areas of open country is managed and the extent to which vehicles are controlled or prohibited. The suggested policy that should apply is considered in paragraph 8.27.

(viii) *anchorages, jetties and landing places* which permit access to the shore from the sea. It will be necessary to ensure, so far as this is possible, that the number of persons who land at these points is consistent with the policy for the management zone in question. This aspect is considered further in paragraph 8.31.

Vehicular access

8.3 The policy governing vehicular access in a Heritage Coast should include these matters:*

(a) Road capacities and widths should be related to the acceptable capacities and levels of use of the resources to which they give access (see paragraphs 6.5 to 6.9).

(b) The provision of parking spaces should also be related to the acceptable aggregate capacity, and all parking should be in places provided for the purpose. Special measures should be introduced to avoid the obstruction of narrow lanes and tracks which might be used by rescue vehicles.

(c) All through traffic and traffic not requiring access to the shore or to heritage features or any of the facilities provided (except for service vehicles and vehicles of local residents) should, if practicable, be discouraged. (This may be achieved by means of diversions, the re-design of intersections, and the careful wording of signs and notices.)

(d) Parking should be severely restricted or prohibited at any points likely to endanger the character, solitude, scientific interest or wild life of the Heritage Coast (whether resulting from the presence of the vehicles themselves or their occupants). Vehicles should not be permitted to be driven or parked on beaches.

* Whilst it may be felt that these principles are stating the obvious, no apology is made for including them because experience suggests that many local authorities pay insufficient regard to them. In a Heritage Coast it is particularly important that principles such as these should be applied.

(e) In remote zones, access should be limited to essential service vehicles and those requiring access to farms and isolated buildings such as lighthouses and coastguard stations. Measures should be introduced to discourage, and where necessary prevent, the entry of all other vehicles.

(f) In remote zones the construction of new roads or the upgrading of tracks to motorable standard should be resisted unless it can be shown that such proposals will not significantly increase the number of vehicles regarded as non-essential in (e) above.

(g) In transitional zones steps should be taken to restrict or prohibit vehicular access to those places where damage to the habitat can be done only too easily.

(h) Where the number of vehicles in a transitional zone already endangers the character of the coast or causes erosion or other damage which cannot reasonably be remedied, existing parking facilities should be re-sited further from the area of conflict. This may mean the stopping-up of a road a few hundred yards short of the area in question and allowing access beyond this point only to those on foot. This would apply particularly to roads which lead directly onto a beach or into the heart of a sand dune system.

(i) Where the limited width and special character of a road within a Heritage Coast justifies it, a one-way system or restrictions on the size, weight, speed or type of vehicles (e.g. coaches and heavy goods vehicles) should be introduced.

(j) In intensive zones attention should be given to the provision of roads and parking places which offer convenient access to the principal beaches and features of interest.

8.4   Careful control of vehicular access and the management of traffic are basic to the concept of Heritage Coasts, because uncontrolled numbers of vehicles can and do cause considerable damage to the environment. This arises from the visual effect of the vehicles themselves, excessive noise, the erosion and damage caused to vulnerable habitats, and above all from the large numbers of persons the vehicles take into the area. To provide uncontrolled vehicular access to the greater part of Heritage Coasts in response to increased demand is not likely to lead to their wisest use. Completely motorless zones would occupy relatively small portions of any particular Heritage Coast, and in some there might not be any. At peak times many main roads near or on the coast are congested but the minor roads and lanes are relatively unused. Where this is so, additional traffic may be accommodated by the increased use of these lesser roads without any new construction taking place. The measures envisaged to restrict traffic altogether in remote zones and to control vehicles elsewhere in a Heritage Coast, may cause some initial public opposition and, if accepted, lead to a gradual change in social habits (see also paragraph 6.14). The services of information and interpretation must, therefore, put across in an effective way the reasoning behind any measures of management that appear restrictive.

8.5   The physical management of traffic in the ways described may be assisted by methods such as those set out in paragraph 6.13. Much may be achieved by encouragement and persuasion. For example, motorists may sometimes be channelled along certain suitable routes simply by means of carefully worded notices and signs which indicate where vehicles are welcome rather than where they are prohibited. Where this is not successful the aim should be to discourage, rather than to prohibit, the use of other less suitable routes, and only where this proves ineffective should stronger measures be adopted. Close co-operation between local planning authorities and local highway authorities will always be necessary.

8.6   Where the access policy affects private roads and motorable tracks, every effort should be made by local planning authorities to secure the agreement of the farmers or landowners, and to work closely with them, in controlling the use of these routes by the public.

8.7   Where a road is a public highway, however, special traffic regulation measures will be necessary. Section 32 of the Countryside Act 1968 provides a basis whereby authorities may regulate, restrict or prohibit traffic in the interest of preserving or enhancing beauty or affording wider opportunities for recreational enjoyment within National Parks, Areas of Outstanding Natural Beauty, country parks, nature reserves, National Trust land, or along long distance footpaths. The measures are exercised by the appropriate Minister on our recommendation. For the most part Heritage Coasts are likely to coincide with these areas but notwithstanding the extended powers provided by s.126 of the Transport Act 1968, it is considered that such provisions should also apply to all land within Heritage Coasts. In doing so, powers to secure traffic management would be available within those parts of Heritage Coasts not so far designated as National Parks or Areas of Outstanding Natural Beauty.

8.8   If part of a Heritage Coast is suitable for definition as a remote zone but it includes a motorable road or track, there may be a case for prohibiting the use of the road or limiting its use to essential vehicles.*

Parking facilities

8.9   As already suggested, the siting and capacity of parking facilities is an important factor in determining the desired pattern of access. Parking places must be provided at key points to serve specific resources or features of interest. Where possible, parking facilities should be designed for the particular purpose for which motorists can be expected to park (for example, to picnic, or to gain access to a beach, or to enjoy the scenery from a prominent viewpoint), and although these may occasionally overlap, the layout and capacity of the parking place should normally reflect this main function.

8.10   The different types of parking place within a Heritage Coast may include:

   (i) *Settlement car parks*, sited centrally to serve visitors to the villages and towns within the Heritage Coast rather than visitors to the shore facilities. These car parks should be outside the three principal zones.

   (ii) *Roadhead car parks* at road terminals, often sited a short distance back from the beach, viewpoint or other feature to which the road leads. Public footpaths should radiate from these car parks to the beaches or other features; if distance is considerable a mini-bus service might be provided. The car parks themselves could well be sited on the line of the inland alternative to the cliff or shore footpath (see paragraph 8.16 (f)).

   (iii) *Beach car parks* to serve the principal beaches and provide all-day or long-term parking. These should be sited close to, but not on, the beach itself. In some cases the pressure arising from excessive numbers may justify roadhead

---

* See also paragraph 8.25 regarding the re-classification of 'roads used as public paths'.

treatment (as in (ii) above) instead. Special parking spaces adjacent to the beach may be necessary for the elderly or incapacitated.

(iv) *Viewpoint car parks* to cater for relatively short-term parking at cliff edges, hilltops and other viewpoints. The car parks should be one of two types: those near the viewpoint but necessitating a short stroll to the actual vantage point, and others so sited that the view may be enjoyed from the car. Great care will be needed in the siting and design of both types; the first type must not mar the quality of the view itself when seen from the viewpoint and the second type, having regard to its more prominent position, must not be so conspicuous as to give rise to too large an expanse of glass and metal when viewed from below or from the distance.

(v) *Picnic area car parks*, in which the parking spaces should form an integral part of the design of the picnic place.

(vi) *Car parks associated with features of special interest* such as monuments, antiquities and other features. Like viewpoint car parks, these should be one of two types, either adjoining the feature and offering direct access to it, or sited some short distance from it with footpath access. The choice of the appropriate type will depend on the nature of the site and the degree to which its enjoyment may be marred by the proximity of vehicles.

(vii) *Vista car parks*, envisaged in connection with scenic roads designed principally for pleasure driving (see paragraph 8.13).

8.11   All car parks must be located with special care so that management zone policies are not frustrated. Scope may exist for collaboration between landowners and others in the provision of car parks as part of a joint management scheme. The location and capacity of parking areas near the boundaries of remote zones will need to be carefully considered to avoid saturating these zones. It should normally be possible for those seeking access to remote zones to leave their vehicles in a park of one of the above types sited within a transitional zone, and reasonably close to the boundary of the remote zone. In some instances however special car parks may be desirable nearer the boundary.

8.12   Peak demand for parking may well exceed the capacity of permanent facilities, and to enlarge these would involve unreasonable expense. One solution may be to provide temporary parking places; for instance, farmers and local residents may be willing to allow motorists to park on private land. Care will, however, be needed to ensure that this is a temporary arrangement, otherwise the capacity outside peak periods may exceed the number of vehicles the Heritage Coast could safely contain without damaging its character. Local people may not be prepared to provide this service unless encouraged to do so. Local planning authorities should therefore consider making informal agreements with landowners to provide any necessary facilities at specific locations, and to prescribe such details as the number of vehicles permitted, access arrangements, and so on. The Special Development Order referred to in paragraph 7.12 would give local planning authorities control over temporary parking.

Scenic roads

8.13   Scope may exist within the transitional zones of some Heritage Coasts for the creation of a road designed primarily for those for whom sightseeing from a car and driving for pleasure provide a major source of enjoyment. Sometimes such a road could be created comparatively cheaply by the improvement of existing minor roads and tracks (e.g. forestry roads, with the co-operation of the Forestry Commission) and the construction of a few short connecting lengths. Extensive provision of scenic roads of this kind within Heritage Coasts would usually be unwise.

8.14　Scenic roads would not cater for through traffic and should in fact be designed to discourage it. Methods by which this may be achieved, in appropriate circumstances, may include the provision of a rough surface, or a limited width, or an irregular or winding alignment. It may be desirable to impose a speed limit, and, during peak holiday periods, one-way and toll systems of operation may be justified. Ample lay-bys, in the form of specially-designed vista car parks, would be needed at strategic points where motorists could enjoy the view for a few moments before driving on to the next stopping place. Long-term parking at these points should not be encouraged, but viewpoint car parks (and picnic sites) off the road itself should form part of the scheme. Scenic roads would enable the pleasures of driving to be enjoyed without the attendant delays, congestion and danger that are caused by heavy commercial traffic or fast through traffic on other roads.

Footpaths

8.15　The view is sometimes expressed that in an increasingly motorised age a footpath system has little part to play except for providing facilities for a small section of the community who enjoy walking as a recreation in itself. But this overlooks the need to cater for the movement of visitors on foot to and from the car parks, and lead them conveniently and safely to beaches and other facilities, some of which will need to be approached on foot for at least part of the way if their essential character or scientific interest is to be retained. Furthermore the management of a Heritage Coast calls for measures to regulate the intensities of use and activity, and the routeing of footpaths is an effective way of achieving this. Remote zones, for instance, would be approached either directly from the sea, or by means of footpaths and bridleways alone. Most keen walkers are also car owners and careful attention should be given to the siting of car parks in relation to the footpath system—for example, by providing the means of returning to the car by a different route. Also many family motorists will welcome the opportunity to park the car and take a walk, provided a well-marked and convenient path exists, and that they know where it leads. They are, however, understandably deterred by the absence of signposts or display maps, and by paths which are overgrown or lead nowhere in particular.

8.16　The following is suggested as the footpath policy that should apply within a Heritage Coast:

(a) The footpaths should provide a convenient system for walkers between focal points, as an alternative to vehicular roads. The principal focal points that should each be linked to one another by the network include villages and settlements, car parks, beaches, hilltops, headlands, viewpoints, picnic places, railway stations, bus and coach terminals, principal local bus stops, isolated hotels, camping sites, youth hostels, and features of special interest such as lighthouses, antiquities and monuments. Some of these focal points will be of greater significance than others, and a hierarchy of paths would be preferable. The paths should link with those in the remainder of the local planning authority's area.

(b) The paths should divert walkers away from fragile habitats and lead them to specific areas, in accordance with the management policy for the zone concerned.

(c) Since an ideal footpath network would aim to segregate pedestrians from vehicles, the need for walkers to follow a road for more than a short distance in order to reach the next section of path should be reduced to the minimum.

(d) Where a path ends abruptly at a road, but not near a focal point, a new link path should be provided to connect with the nearest focal point or with another path.

(e) A footpath should be provided along each Heritage Coast parallel to the shore, either adjacent to it or at cliff top level. This would not only facilitate access to cliff tops and to the smaller beaches and coves not directly served by roads but should also offer a direct and convenient route on foot to the next resort, beach or focal point along the coast. At intervals a spur path, or steps, should be provided to permit access to the water's edge. In places the special vulnerability and importance of a particular habitat, or other circumstances, may cause a minor diversion inland. In some instances the Heritage Coast may already include a national long distance footpath established or proposed under s.51 of the National Parks and Access to the Countryside Act 1949.

(f) In addition to a path along the cliff or shore, the footpath system should, if possible, offer alternative return routes. The need to retrace one's steps along the same ground often deters visitors from making use of coastal footpaths. A return route may be provided by means of loops on the cliff path, connecting with it at intervals, or by means of a separate parallel path a short distance inland. Either would be an acceptable alternative for the return journey, and could take advantage of natural features such as a hill ridge or a river. An indented coastline with a series of headlands presents an opportunity for one path along the cliff edge and another by means of links across the neck of each headland. It would also sometimes be possible for alternative paths to serve as links between some of the focal points and features described in (a) above, for example, to connect roadhead car parks $\frac{1}{4}$ mile or so inland.

(g) By means of new paths or connecting links, routes should be provided along the lines of the main interesting natural features, particularly hill ridges, valleys and rivers. Advantage should also be taken, if possible, of the existence of other linear features such as canals and disused railways.

(h) Special care should be taken to ensure that the signposting of all paths is adequate, and yet simple and unobtrusive. It should be supplemented where necessary by display maps and simple way-marking. In remote zones all signposts should be replaced by unobtrusive waymarks in order to preserve the landscape in its natural state. Elsewhere throughout the Heritage Coast, in those cases where the standard footpath signpost prescribed in the Traffic Signs Regulations 1964 (S.I. 1857) is considered inappropriate, advantage should be taken of the Traffic Signs Amendment Regulations 1966 No. 490 which allow the use of smaller and less obtrusive signposts. Except at focal points and the principal path or road intersections (where a colour code might be used to identify particular routes shown on a display map) way-marking should be kept to a minimum, sufficient only to guide walkers on obscure or dangerous stretches of path in open country or near cliff edges and bogs. Within remote zones such marking should be carried out only in exceptional circumstances.

(i) Adequate informative and interpretative techniques should be used to ensure that the existence of the footpath network is made known to visitors to the Heritage Coast. Visitors should be able to choose from several possible routes. These might vary from an hour's stroll to a full day's walk. It should be possible for a visitor readily to discover, before setting off, what alternatives exist, how long they will take to follow, the type of surface likely to be encountered, whether any special care (or stout footwear) is called for, and some assurance that the path is sufficiently well marked.

8.17 The provision of a footpath network of the kind described implies the full use of the powers contained in the Highways Act 1959, the Countryside Act 1968, and the Town and Country Planning Acts 1962 and 1968 to create, divert or close public footpaths. The statutes define the purposes for which orders may be made: these include, for example, the provision of a more commodious path or the securing of a more efficient arrangement of land use. These powers can be used variously by highway authorities, local planning authorities, and also in certain circumstances by the Ministry of Transport, the Ministry of Housing and Local Government and magistrates' courts. Close co-ordination between all these is essential if the footpath network is to fulfil its role effectively in the management of a Heritage Coast. Close liaison with landowners and farmers will enable management techniques to be introduced to minimise the possibility of any conflict with agricultural interests—for example, by the provision of self-closing gates on footpaths, or the replacement of farm gates by cattle grids. Measures to achieve this liaison are suggested in Chapter 12.

8.18 In many Heritage Coasts a substantial portion of the network envisaged is likely to exist already, and all that may be required will be a few connecting links, and a comprehensive scheme of signposting and marking paths to promote their use. The Countryside Act 1968 enables rambling clubs and other local voluntary organisations to play a part in signposting and marking paths, with the consent of the highway authority, and every opportunity should be taken to encourage this. Voluntary assistance may also be forthcoming for the surveying of routes.

8.19 Paths of the kind referred to in paragraph 8.16(e) offer exceptional opportunities for the enjoyment of the coastal scenery. As Heritage Coasts would represent the best examples of this scenery in England and Wales such paths within them would be of considerable national significance. The establishment of a long distance route under s.51 of the National Parks and Access to the Countryside Act 1949, is proof that the national importance of a path which permits extensive journeys to be made has been recognised. Thus Exchequer grants of 100% are payable to local authorities in respect of the costs of creating and maintaining such routes. A criterion for establishing a long distance route is that it should permit an extensive journey to be made. In practice this has normally been regarded as one of not less than 50 miles, which requires at least one overnight stop, in an area of high quality scenery.

8.20 In 19 of the 34 areas suggested as Heritage Coasts an approved long distance route exists, in most cases as a cliff or shore path of the type suggested in paragraph 8.16(e). In the remaining Heritage Coasts, there are, with very few exceptions, no proposals for a long distance route under the 1949 Act. Moreover the length of the

Heritage Coast (or the proximity of adjacent areas of Heritage Coast which it could link) would seldom justify such a route. In these circumstances we consider that the national significance of a path along the cliff or shore of a Heritage Coast should be recognised by the payment to local authorities of an Exchequer grant for its creation and maintenance as if the path were part of a long distance route established under the 1949 Act.

8.21 Existing footpaths in coastal areas frequently become overgrown and discourage or impede progress along them. The cheapest and simplest way of keeping paths open and unobstructed is to encourage their regular use. But unless they can be seen by visitors to form part of a planned network, some paths are never likely to be used sufficiently to keep them in good condition. This emphasises the need for adequate promotional measures and signposting, as well as clearance of undergrowth, nettles and brambles and the maintenance of stiles, gates and footbridges. Much of this will in any case be necessary from time to time even where paths are well used. Here, too, there is scope for using voluntary labour which may well be forthcoming from rambling clubs and outdoor bodies. This direct participation would foster a genuine interest in, and an increased use of, the paths of a particular locality, and would also lessen maintenance costs. The inclusion of the less-used paths within the regular walking programme of local clubs and societies, especially outside the main holiday season, would also help to keep them open and unobstructed. With encouragement, such bodies may be prepared to 'adopt' a particular stretch of coast, or even an entire Heritage Coast, and thereby keep the paths under constant review. Local authorities should be encouraged to explore these possibilities.

8.22 Concern is sometimes expressed at the concept of a rationalised footpath system, or even a carefully planned network of the kind envisaged in the White Paper on *Leisure in the Countryside* (Cmnd. 2928). The objection appears to be that this would introduce an element of artificiality into a footpath system, much of the charm and attraction of which is its casual unplanned nature. What is often not realised is that a large part of the network required often exists already and includes many attractive and long-established paths which have evolved slowly over the centuries, including some of those paths formerly used by coastguards. Often relatively few new links are required to make the system fully serve its purpose, and increased use would help considerably in maintaining any path in a walkable condition. It need not follow that all other paths not having a 'trunk' role in the network should be closed. But the justification for retaining some minor paths of limited length which connect neither with main features nor with other paths in the area is questionable. The original need for many of these paths has long since disappeared and some are rarely, if ever, used. They are often seriously overgrown, of inconvenient alignment, and necessitate much road-walking to reach them. Some of these could well be extinguished without any appreciable loss, in return for some of the new paths and links suggested in the preceding paragraphs. With care it should be possible to create a network which fulfils an important role without detracting from its enjoyment by those who appreciate the casual and unplanned nature of many of the older footpaths.

Bridleways

8.23 The policies governing bridleways and footpaths in a Heritage Coast should be closely linked. Since walkers, as well as horse riders and cyclists, are entitled to use bridleways many of the points made in connection with footpaths will apply. The specialised nature and requirements of bridleways will however dictate a much less intensive network than that envisaged for walkers. The pattern should consist of trunk routes within the footpath network, and especially those which provide links between settlements, beaches, picnic places and viewpoints. In those parts of the country where there is likely to be a demand for riding holidays, isolated hotels and youth hostels should be similarly linked. Stables would constitute the principal focal points, and car parks could be avoided. It should be possible for riders to make reasonably long and continuous journeys on bridleways without having to follow extensive stretches of road. Provision should also be made for the routes to extend beyond the boundaries of the Heritage Coast. Special signs, distinct from those used to mark footpaths, should be designed to enable the routes to be followed clearly. Facilities such as drinking troughs and hitching posts may also need to be provided.

8.24   Attention should be given to the effects of erosion arising from hooves, and areas likely to suffer special damage in this way should be avoided. The consequences of providing access within remote zones for horse riders and cyclists should be carefully considered. Some remote zones may have to be confined to walkers.

8.25   We envisage that the creation of a bridleway network would be effected, as in the case of footpaths, by making new links or by the diversion of existing ones. There is the possibility of upgrading paths to bridleways (i.e., by closing the path and creating a new bridleway along the same line). Other opportunities may arise from the special review and re-classification of 'roads used as public paths', for which provision is made in s.31 and Schedule 3 (Part III) of the Countryside Act 1968. Where the test for re-classification (Schedule 3(10)) cannot justify vehicular use, local authorities whose areas include Heritage Coasts should be encouraged to take advantage of this review in re-classifying roads as bridleways and thereby providing strategic links required in the network. Byelaws which prohibit cyclists from using specified bridleways outside villages should be reviewed in the light of the need to provide adequate routes for pedal cyclists within a Heritage Coast.

8.26   Close consultation with landowners, farmers and local interests, including riding clubs and other establishments whose members are likely to constitute the chief users of these routes, is desirable. By this means, management techniques of mutual benefit may be secured. These may perhaps lead to the re-siting of cattle grids for the convenience of riders and the provision, on bridleways, of self-closing gates and those which may be opened by riders without needing to dismount.

Access to open country

8.27   Wherever possible, access to uncultivated land of special recreational value immediately adjoining the shore should be made available to the public. Open country in this context includes beaches, river banks, cliff tops, headlands, sand dunes, downland, scrub, heath, commons, rough pasture, and other uncultivated land where random access, either legal or *de facto*, is permitted. Within some Heritage Coasts considerable stretches of land falling within this definition are already publicly owned or held inalienably by the National Trust, or are registered as common land, or are otherwise available for public access. Problems have however arisen: in some parts open land has been ploughed up to the cliff edge, with the result that the public are excluded, and elsewhere landowners have denied access by barriers, barbed wire and warning notices.

8.28   Local planning authorities must choose the appropriate remedy carefully. The establishment, as a right of way, of the path suggested in paragraph 8.16(e) may go some way to provide limited public access, but an access agreement or order, or even acquisition, may be necessary to allow full recreational use of the open country in question. It is important that a local planning authority should be in a position to co-ordinate the management policies which apply to the various stretches of open country included in a Heritage Coast, and for this purpose close consultation with landowners and public bodies will be required. Much may be achieved informally in an atmosphere of goodwill. In practice we envisage that the policy outlined in paragraph 8.27 would be achieved by a combination of various methods along different stretches of coast, with preference for the use of agreements. In our main report* we commend the use of management agreements with positive convenants to secure such goals.

8.29   Mention has already been made of the problem whereby open land is ploughed up to the cliff edge to discourage or curtail public access. The scale of this operation varies widely but it is causing concern in parts of the South West Peninsula. The cliff land is mostly of little agricultural value, whereas its recreational importance, for picnicking and walking and for the exceptional coastal views that may be obtained from it, is immense; undoubtedly it justifies measures as strong as those which apply to moorland and heath in a National Park, by virtue of s.14 of the Countryside Act 1968. This section, which may by Ministerial order apply to any moorland or heath within a National Park which has not been agricultural land (other than rough grazing) during the preceding 20 years, requires an occupier to give six months' written notice to the local planning authority of his intention to plough or otherwise convert it to agricultural land. This enables the local planning authority to consider, in cases where the informal approach has not succeeded, whether acquisition, or an access order or agreement, should be initiated. This section would thus apply, if the Minister sees fit, to parts of those 6 of the 34 suggested Heritage Coasts which fall within National Parks. Having regard to the special significance of this marginal land in other Heritage Coasts where the coastal scenery is also of national importance, it is considered that the provisions of section 14 should be extended so as to be capable of application to any stretch of heath or moorland within a Heritage Coast.

8.30   Exceptions to the general policy of providing access suggested in paragraph 8.27 may be justified in those places (e.g. dunes or wetlands) where the presence of a large number of people not confined to a specific path could cause damage. To some extent the absolute numbers may be controlled by the careful siting of car parks and the regulation of boat landings, but it may still be necessary to channel visitors along defined routes in places where random access would be harmful. It is possible to achieve this without resorting to fencing or prohibitive notices by the provision, in selected areas, of seats, picnic spaces or defined paths (which may be composed of gravel, pebbles, stone chips or wooden rafting, etc., depending on the nature of the surface) and inviting visitors to use these. Experience has shown that public respect for such arrangements is greatly enhanced by the provision of simple signs which explain clearly the purpose for which this form of management is necessary.

Access from the sea

8.31   The policy with regard to anchorages, jetties and landing places should ensure that the degree of access which may be possible from the sea is consistent with the management policy for the zone in question. The character and solitude of a remote

---

* *The Planning of the Coastline*, paragraph 218.

zone, for instance, would be likely to suffer from the influx of a large number of passengers landing from a motor boat on its shore. Careful management of boat landings throughout a Heritage Coast is indeed essential if the measures described in this chapter to control access and intensities of use are to be meaningful. Legally a general public right of navigation exists over all tidal waterways, and this right may only be extinguished in pursuance of statutory powers. Section 76 of the Public Health Act 1961 enables local authorities to regulate the speed of pleasure boats over the foreshore (and for a distance of 1,000 yards seawards from low water mark) and to impose other restrictions to prevent their navigation in a dangerous manner, or ' . . . without due care and attention or without reasonable consideration for other persons'. Section 249 of the Local Government Act 1933 also enables byelaws to be made for good rule and the suppression of nuisances. In addition, under s.82 of the Public Health Acts Amendment Act 1907, a local authority may make byelaws over the beach or foreshore to regulate public use: the consent of the Board of Trade must be obtained for byelaws affecting the foreshore, as also must the consent of the Crown Estate Commissioners in the case of foreshore in Crown ownership.*

8.32 Local authorities will therefore need to consider carefully which measures to adopt, and under which enactment, in order best to achieve the effective management of the Heritage Coast as a whole. Byelaws may be desirable to confine direct access to selected points and to specify the type of vessel, permitted noise levels, and so on, within a prescribed distance of a remote zone or a popular beach. Measures such as vessel registration, a permit system and the levying of discriminatory mooring charges may all have a part to play. Some form of zoning may also be required to confine powered vessels to specified waters and ensure they do not penetrate areas used principally by sailing boats and smaller craft. For many of these purposes statutory powers may be needed. These aspects, and the determination of those types of boating activity which are acceptable, should be considered in relation to the Chart of Acceptable Activities and the principles suggested in Chapter 7.

---

* So far as Crown foreshore is concerned it is the Commissioners' policy wherever possible to provide for local control in the public interest by encouraging local authorities to take regulating leases of Crown foreshore within their district. This is a quick and effective means of providing control over uses of open foreshore and much of the foreshore is controlled in this way.

1. *North Northumberland:* Bamburgh Castle and sands.

2. *North Yorkshire:* Robin Hood's Bay and Ness Point.

3. *Flamborough Head:* The lighthouse and Selwicks Bay.

4. *Spurn Head:* looking towards Kilnsea.

5. *North Norfolk:*
Morston Salt Marshes.
looking towards
Blakeney Point.

6. *Suffolk:* Cliffs at
Dunwich, looking
south to Dunwich
Common and
Minsmere.

7. *South Foreland:*
The 'white cliffs of
Dover', between
Dover and St.
Margaret's Bay.

8. *Seven Sisters and Beachy Head:* looking towards Birling Gap.

9. *Isle of Wight:* Alum Bay, The Needles, and Compton Bay.

CHAPTER 9.
PROVISION OF RECREATIONAL FACILITIES

9.1   In Chapter 7 a policy is suggested whereby the acceptability of various forms of land use and development within a Heritage Coast may be tested. For recreational activities the implications of this policy are that supporting facilities should only be provided or accepted where the activity itself is an acceptable one within the zone concerned. By means of a clear policy of acceptability on the lines of the model Chart of Acceptable Activities (see Chapter 7) a local planning authority should be able to determine the justification of facilities such as a golf clubhouse, a jetty for the mooring of motor boats, a riding centre or an equipment store for a canoe club. The present chapter outlines a policy which seeks to ensure, firstly, that adequate facilities are available, and secondly that these facilities are so designed and located that the character of the Heritage Coast does not suffer as a result.

9.2   The special quality of the landscape within a Heritage Coast demands that only a minimum of facilities should be provided. These should be sufficient in number only to enable the quality of the scenery, and those activities based on it, to be fully enjoyed. Ancillary attractions such as amusements and entertainments should be confined to resorts or to other settlements outside the boundaries of a Heritage Coast. Within the boundaries most of the facilities should be channelled into intensive zones, leaving remote zones free of all facilities but the natural resources themselves. The greatest sensitivity in siting and design will be necessary, and the criteria outlined in paragraph 7.4 should be applied. In some parts special opportunities may exist for adapting existing buildings such as martello towers, disused lighthouses or coastguard lookouts for use as information centres, shelters or restaurants. The size of car parks and similar facilities should relate to the optimum capacity of the Heritage Coast determined in accordance with the principles outlined in Chapter 6.

9.3   In order to minimise the effect on the landscape of car parks or lavatories and other buildings, it is important that these should be deliberately grouped at a number of principal focal points and should not be scattered throughout the Heritage Coast. A focal point would be any single feature or locality where visitors are likely to pause or spend more time (perhaps even most of their time) whilst in the area. Most focal points would be related to a specific feature such as a beach or a major headland, but any point could combine two or more features—thus a cliff top with a lighthouse would serve both as a viewpoint and a feature of interest and could also be provided with facilities to serve as a picnic area. Most focal points would require a car park of one of the types referred to in paragraph 8.10, although a proportion of individual

*D

features (including isolated viewpoints, minor beaches in transitional zones and features of special interest) should be rendered accessible only by footpath from a nearby car park.

9.4   We consider that the principal focal points within a Heritage Coast would be:

(a) *Settlements*, including smaller villages and groups of buildings either with or without a coastal frontage, and outside the three main zones, but within the Heritage Coast boundary.

(b) *Beaches*, apart from some of the smaller beaches in transitional zones which would be more suitably approached by footpath.

(c) *Roadheads*, from which there is access by footpath to the smaller beaches, viewpoints, features of interest and other stretches, and sited between about ¼ and ¾ mile inland from the shore.

(d) *Main headlands or viewpoints*, other than those reached by footpath from a nearby car park.

(e) *Features of special interest*, other than those reached by footpath from a nearby car park.

(f) *Picnic areas*, comprising specially laid out spaces with ancillary facilities.

(g) *Anchorages*, consisting of a small self-contained complex of landing and safety facilities.

9.5   The determination of appropriate facilities to be provided at focal points should be based partly on the management zone (which would determine the scale and capacity) and partly on the nature of the focal point. The model Facility Provision Chart (see page 45) illustrates the scale and range of facilities envisaged. A more sophisticated approach based on the individual resource characteristics of the particular Heritage Coast would be desirable, but the Chart should be sufficient to indicate approximately what we envisage.

9.6   The following are the sixteen main types of facility (as shown on the Facility Provision Chart) appropriate to a Heritage Coast:

(i) *Car parks*, of the types described in paragraph 8.10, to serve settlements, beaches, roadheads, viewpoints, anchorages, picnic areas and features of interest;

(ii) *Jetties and landing facilities*, adequate to meet the needs of visitors in relatively small numbers who wish to practise sailing, power boating, canoeing or rowing;

(iii) *Moorings*, providing simple facilities only, and on a much reduced scale, for not more than two or three vessels;

(iv) *Fishing jetties*;

(v) *First Aid Posts*, intended chiefly for the main beaches;

(vi) *Lifesaving equipment*, which could be stored at a First Aid Post or in unattended boxes containing lifelines, etc., on beaches or below cliffs;

42

(vii) *Lavatories*;

(viii) *Stables*, offering facilities for horse riders and pony trekkers but not necessarily providing accommodation (see para. 9.8);

(ix) *Shelters* or windbreaks at the main beaches or at exposed locations away from car parks;

(x) *Litter Bins*;

(xi) *Restaurants*;

(xii) *Cafes*, offering basic catering facilities similar to (xi) but on a much reduced scale;

(xiii) *Picnic spaces and facilities*, including a defined picnic area with tables and benches or logs for seats, a water standpipe and tap with drain, a concreted area for fires and picnic stoves, a shelter or windbreak, and litter receptacles;

(xiv) *Information Centres*, serving the whole or a large part of a Heritage Coast and the area adjoining, offering guide books, leaflets, maps and other publications, and staffed by persons able to advise visitors on features of interest, local walks, accommodation and refreshment places, and so on;

(xv) *Information Posts*, offering facilities on a much reduced scale by means of manned or unmanned kiosks, or self-service dispensers containing leaflets, relating mainly to a specific feature or locality within a Heritage Coast;

(xvi) *Observation Platforms* offering uninterrupted views, for example at dangerous cliff edges and at places where trees would otherwise obscure the view from ground level.

9.7   We consider that the local planning authority for a Heritage Coast should adopt and incorporate into its development plan measures such as those described in this chapter in order to determine which facilities are appropriate at any particular site. This will serve as a basis both for assessing planning applications for new proposals and for determining which facilities should be provided by the local authority. An examination of the existing facilities within the area will indicate how far they already meet the requirements. One of the aims of the management policy should be to determine which of the existing facilities should, ultimately, be re-sited and grouped at principal focal points. Close contact with local landowners, public agencies and other bodies is essential. By this means, an individual site owner would be able to determine the nature and location of facilities likely to be approved and would be discouraged from submitting applications for proposals not in accordance with the policy. It may also be possible to arrange a wider use of facilities hitherto reserved for a particular group of users. By means of agreements involving some financial consideration, farmers, hoteliers and others may be persuaded to make facilities such as lavatories or parking areas available to the general public, perhaps at peak times. (See also paragraph 8.12.) These measures would keep the number of facilities provided to a minimum consistent with the landscape of the Heritage Coast.

9.8   It has already been suggested that overnight accommodation for visitors would not normally be considered appropriate in a Heritage Coast. In accordance with the code described in Chapter 7, new buildings such as hotels, flats or chalets, and new caravan sites, would be confined to existing settlements or areas outside a Heritage Coast. This principle would also apply to any *new buildings* for field study centres, outdoor pursuits bases, riding centres, or youth hostels, in spite of the fact that the majority of their users would probably spend most of their time in activities entirely acceptable in a Heritage Coast, and even though the need for car parking space may be limited or non-existent.

9.9 It may however be possible to provide some accommodation for visitors by means of the conversion or adaptation of existing buildings. If this would lead to the economic use of a building which enhances the landscape or character of the area and would otherwise fall into disrepair, such a change of use may be regarded as beneficial and should be welcomed. Great care would however be needed if any material modifications to the exterior proved necessary. Occupiers of existing farm-houses or cottages may, in addition, be prepared to provide bed and breakfast facilities.

9.10 Motorised camping would normally be unacceptable in most parts of a Heritage Coast. Where it already exists regard should be had to its effect on the area. Light-weight camping, involving smaller numbers and the use of smaller tents, and not requiring on-site car parking facilities, would be generally acceptable on a limited scale. The Special Development Order we recommend in paragraph 7.12 would enable the local planning authority to secure complete control of camping.

FACILITY PROVISION CHART

Zone	Principal Focal Point	Facilities normally provided at each site	Additional facilities provided at selected sites only
Settlements	Settlements, either with or without a coastal frontage	Settlement car park Beach car park Information Centre (or Post) First Aid Post Restaurant Cafe Lavatories Litter Bins	Stables Jetty and landing facilities (or moorings) Lifesaving equipment Shelter
Intensive	Beaches	Beach car park Lavatories Information Centre Shelter Jetty and landing facilities First Aid Post Lifesaving equipment Restaurant Litter Bins	Cafe Picnic space and facilities
	Roadheads	Roadhead car park Litter Bins	Lavatories Information Post First Aid Post
	Main headlands or viewpoints	Viewpoint car park (either adjacent or with foot access)	Lavatories Information Post Restaurant (or Cafe) Picnic space and facilities Observation Platform Litter Bins
	Features of special interest	Car park (either adjacent or with foot access)	Lavatories Information Post Cafe Litter Bins
	Picnic areas	Picnic space and facilities Picnic area car park Litter Bins	Information Post Lavatories
	Anchorages	Car park Jetty and landing facilities Lavatories First Aid Post Lifesaving equipment Fishing jetty Litter Bins	Information Post Restaurant

Zone	Principal Focal Point	Facilities normally provided at each site	Additional facilities provided at selected sites only
Intensive *(Contd.)*	Individual locations (other than those listed above) where justified		Moorings Fishing jetty Lifesaving equipment Vista car park (on a scenic road) Litter Bins
Transitional	Main beaches	Beach car park Lavatories Litter Bins	Information Centre (or Post) Jetty and landing facilities (or moorings) First Aid Post Lifesaving equipment Fishing jetty Restaurant Cafe Picnic space and facilities Shelter
	Smaller beaches	(Served by roadhead facilities)	Moorings Fishing jetty Litter Bins
	Roadheads	Roadhead car park Litter Bins	Lavatories Information Post
	Main headlands or viewpoints	Viewpoint car park (either adjacent or with foot access)	Cafe Observation Platform Litter Bins
	Features of special interest	Car park (either adjacent or with foot access)	Litter Bins Lavatories
	Picnic areas	Picnic space and facilities Picnic area car park Litter Bins	Information Post Lavatories
	Anchorages	Car park Jetty and landing facilities Fishing jetty Litter Bins	Lavatories Information Post First Aid Post Lifesaving equipment Restaurant
	Individual locations (other than those listed above) where justified		Vista car park (on a scenic road)

CHAPTER 10.
LANDSCAPE IMPROVEMENT

10.1 The need to conserve the quality of the landscape should be implicit in all aspects of Heritage Coast management. How far this is achieved will depend largely on the sensitivity with which individual policies relating to access, development control, the provision of recreational facilities, and so on, are implemented. Along some stretches of coast additional measures will be necessary. These include restoration, landscaping, tree planting, and the removal of disfigurements.

10.2 Schemes within Heritage Coasts to remove disfigurements which materially detract from the beauty of the landscape are suggested in our main report* as one of the five kinds of improvement scheme which should justify priority treatment on a national basis. These schemes, which would qualify under s.34 (1) (10) of the Countryside Act 1968 for 75% Exchequer grants, should, it is suggested, be included in clearance programmes which local planning authorities would be called upon to submit annually to Ministers. Local planning authorities should be required, as soon as boundaries have been agreed, to carry out a review of each Heritage Coast to decide on which sites action should be taken.

10.3 Every endeavour should be made to co-ordinate management policy between public agencies and the local planning authority concerned with the Heritage Coast. Matters for consideration in this way might include the undergrounding or re-routeing of overhead electricity and telephone lines, the sharing of any necessary masts, and the re-design or removal of obsolete or unnecessary buildings, structures, advertisements and notices.

10.4 Consultations with private landowners may lead to action which would eliminate or reduce the effect of features which disfigure the landscape or obstruct views. The immediate surroundings of many monuments, antiquities and other features of interest could be improved by the regular clearance of undergrowth, the repainting of rusty iron railings and the repair of fences or stone walls. The appearance of unused pieces of land could be improved, and quarries, spoil heaps and mineral workings made less conspicuous, and also safer for children and animals. Much of this could be done by voluntary labour.

10.5 Co-operation with private owners in the clearance of litter from the less accessible parts of the coast is a possibility. Within some of the National Parks, local

---

* *The Planning of the Coastline*, para. 142.

planning authorities have made agreements with farmers and others, whereby, in return for a small annual payment, litter is cleared periodically from those parts which cannot be reached by refuse disposal vehicles. A similar arrangement should be encouraged in the remote zones and the less accessible parts of transitional zones in a Heritage Coast.

10.6 Signs and advertisements present special problems in areas of great beauty. The Special Development Order we recommend would prevent the erection of new signs and notices, but it may take a long time to clear away those which already exist. Landowners and others in a Heritage Coast should be persuaded to keep the number and size of signs and notices to a minimum, and to replace or remove those which are no longer strictly functional.

10.7 New notices should be attractive in style, lettering and design, and make use of pictorial symbols where appropriate. Special care will be needed in the use of signs to direct, restrict or prohibit traffic in the ways described in Chapter 8.* The desirability of a 'house style' for official notices is suggested in the following chapter.

*See also para. 8.16(h) regarding the use of less obtrusive footpath signs.

48

CHAPTER 11.
INFORMATION AND INTERPRETATION

11.1   One of the most important aspects of Heritage Coast management is likely to be the provision of information and interpretative services which aim to promote a closer understanding of, and interest in, the coastal environment. An understanding of an environment leads to a respect for it, and the building up of a body of informed opinion is one of the most effective ways of guaranteeing the conservation of parts of the coast which might be damaged.

11.2   One of the principal aims of management would be to enhance the public enjoyment of the coast. Some stretches of Heritage Coast are not as fully enjoyed as they might be, and many of their resources could bear increased use without any detrimental effects. Visitors often miss seeing some of the best parts of these coasts, either because they are unaware of their existence or because of lack of information about their special qualities. An attractively designed sign or waymarked footpath leading to a viewpoint, an antiquity, or some other feature, may help greatly. By promoting a greater interest in, and thereby a greater use of, these existing resources much may be done to enhance the enjoyment of a Heritage Coast by visitors without the need to provide any new facilities.

11.3   The policy for information and interpretation in a Heritage Coast should:

  (i) emphasise the opportunities that exist for different types of recreation, inform visitors of the most suitable places for these activities, and call attention to those which would lead to a more even distribution of people throughout the area (see paragraph 6.4(6));

  (ii) promote and encourage, in each of the management zones, those activities indicated as category 'A' in the Chart of Acceptable Activities (see Chapter 7);

  (iii) provide information for visitors with specialist interests, indicating, for example, the most suitable areas for geological study, or for ornithology, and the best places from which to take photographs;

  (iv) ensure that the optimum capacity of the Heritage Coast is not exceeded by providing visitors, especially at peak times, with reliable information about the availability of parking spaces, by advising on facilities already fully saturated, and by discouraging visitors from going to areas where there is no room for them;

(v) direct visitors to certain areas and steer them away from the more vulnerable parts, by means of persuasion and encouragement, and explain why such measures are necessary (for example, to make it clear why visitors are asked to keep to the paths in sand dunes, and to explain why access to a particular beach is confined to those on foot);

(vi) promote the greater use of existing resources such as antiquities, viewpoints, and other features of interest, especially the lesser known examples, and including those privately owned;

(vii) make visitors aware of any special dangers that may exist, including dangerous currents, beaches likely to be cut off at high tide, concealed mineshafts, crumbling cliffs and those too friable for climbing;

(viii) exercise the greatest discretion and sensitivity in the design and siting of signs, notices, structures and other interpretative features.

11.4 Although the implementation of this policy would apply throughout a Heritage Coast, regard should be had to the management policy for individual zones. In remote zones, for instance, information should be given by means of printed leaflets or maps, and not by conspicuous signboards or other features on the ground. Those signs which are essential should be sited at a low level.

11.5 The policy we suggest implies a wide variety of methods and techniques to guide, inform and encourage visitors to make the best use of a Heritage Coast and the features it contains. The following are the principal methods envisaged by which this should be achieved:

(a) *Information Centres and Posts*. These should be staffed by people with detailed local knowledge, and should offer a range of maps and other publications as well as advice on all aspects of recreation within the Heritage Coast.* Information Centres should be regarded as the principal sources of tourist information within a Heritage Coast and could well be combined with other buildings and facilities. It may be possible to establish an Information Post in a room of a cottage, or in a shop or restaurant. In certain parts dispensers containing maps or pamphlets, and with 'honesty boxes' attached, may be more appropriate.

(b) *Coastal Centres*. Some of the larger Heritage Coasts may need a greater range of facilities than those provided by an Information Centre. These would include opportunities for field study, and facilities similar to those envisaged in the study centres for which provision is made in s.12 (1) of the Countryside Act 1968 and of which the National Park Centre at Brockhole in the Lake District is an example.

(c) *Warden Service*. We regard the provision of a warden service as essential for a Heritage Coast. In addition to securing compliance with byelaws, wardens would fulfil a vital role in advising visitors on all aspects of the Heritage Coast. In particular, wardens could explain the reasons for managing and conserving the landscape, and the methods used.

(d) *Display Maps*. Mention is made in paragraph 8.16(i) of the need to provide visitors with adequate information on footpaths and alternative walking routes, and to give details of the condition of surfaces and average walking

* See paragraph 9.6 for the distinction envisaged between Information Centres and Information Posts, and see also the Facility Provision Chart for the suggested locations at which these should be provided.

times. Much of this could be presented in the form of a display map installed at each of the main car parks and other key focal points. The map would show how the footpath and bridleway networks relate to the road system, and would recommend routes which lead to key features in the area. The colour codes which relate to any waymarks that may be used should also be shown. Some maps might be orientated and mounted on a horizontal plane.

(e) *Guided Walks.* These would start from Information Centres according to a regular programme at peak times, and at other times by request, and would explain to visitors the main features of physiographical, wild life or general interest in the area. By this means visitors not accustomed to rough or strenuous walking could also be guided along interesting cliff walks and to parts of remote zones they would otherwise be unlikely to visit. The Countryside Unit established on the Pembrokeshire coast by the Countryside Commission, in association with the Field Studies Council, is one example of how this might be done.

(f) *Self-guided Trails.* A wide variety of these is suggested; they may cover a range of different features. See paragraph 11.8 below.

(g) *Panoramic Viewfinders.* Boards should be set up with a diagram or photograph for identifying the main features of the view from selected headlands and viewpoints. Brief interpretative notes should be included.

(h) *Observation Hides.* In localities where special opportunities exist for studying wild life (for example some nature reserves) consideration should be given to the provision of observation hides (with well concealed access) for the use of photographers and others.

(i) *Lectures and Film Shows.* These should be included among the facilities provided at Information Centres and Coastal Centres and would cover aspects of the Heritage Coast and its conservation, including its history, antiquities, folk lore and legend, geology and wild life. The opportunity should be sought to secure the co-operation of local officers engaged in conservation within the area (for example those employed by the Nature Conservancy, the Forestry Commission, the National Trust or local Naturalists' Trusts).

(j) *Exhibitions and Displays.* These, which could also be provided at Information Centres (and with co-operation might be possible in shops and other places), would include displays of local features such as geological specimens, trees, plants and other aspects of wild life, local crafts and manufactures, and books describing the area. Various themes would be possible, including the lifeboat and coastguard services, boating, fishing, trade and literary associations.

(k) *Demonstrations of local activities and crafts.* In addition to static displays, advantage should be taken of any opportunity of demonstrating to visitors aspects of rural life. These might include practical demonstrations of thatching, visits to farms, or conducted tours of coastguard stations, lighthouses, mines and quarries.

(l) *Visitor's Map.* A simple, cheap, specially-prepared map of the Heritage Coast should be produced for sale. (See paragraph 11.10.)

11.6   Neither management nor interpretation would cease abruptly at the Heritage Coast boundary. However, since Heritage Coasts would represent the best examples of coastal scenery in England and Wales there is much to be said for regarding each as a separate entity worthy of interpretation in its own right. The significance of the coast as a unit would be enhanced by the adoption of a distinctive and characteristic symbol for each area (perhaps an animal, or an easily recognisable coastal feature), on the lines of those adopted for the National Parks. This symbol, in association with a 'house style' for use on signs, notices and in publications, would stress the local identity of the area and serve to emphasise that all management measures adopted have the common aim of conserving the area as a whole as part of the coastal heritage.

11.7   The popularity of self-guided nature trails has shown that an increasing number of visitors find enjoyment in following, on foot, short routes recommended by experts, along which attention is directed to specific features of natural history interest. Those features of a trail which attract the visitor (but which are lacking in an ordinary footpath route) include, firstly, an attractive booklet or leaflet to entice the visitor onto the trail (and which also serves as a souvenir), secondly, guarantee of a clearly defined or well-marked path and thirdly, the authenticity of the expert advice offered, which enables an ordinary visitor to identify features, plants, etc., which he would not otherwise notice.

11.8   Heritage Coasts offer opportunities for a variety of trails suitable for a range of tastes. Some trails could have a specialist bias (e.g. geology, or marine biology), others a specific coastal theme (e.g. smuggling) and others could relate to features of general countryside interest (including archaeological and historical sites, buildings of special architectural interest, ancient monuments, locations with literary associations, natural features such as caves and waterfalls, or notable viewpoints). Such routes, extended in scope beyond the original nature trail concept, would be based almost entirely on the footpath network. The principal car parks would form suitable starting and finishing points. Other trails might make use of the cliff or shore footpath. Most trails would require some form of waymarking or identification on the ground (by means of marker posts, etc.) and thus should be confined to the transitional and intensive zones of a Heritage Coast. One example of a typical trail could be included on the reverse of the visitor's map suggested in paragraph 11.10, but it is also envisaged that a separate pamphlet would normally be provided for each trail. These should be available at information centres, shops, cafes or self-service dispensers.

11.9   The adaptation of the trail technique for use by motorists has been little used. In view of the success of nature trails it is considered that a trail which would involve both the use of a car and would require short walks from specified stopping places would be extremely popular. Such a technique should be adopted for use on scenic roads. A motor trail would enable the greater part of a Heritage Coast to be covered in one route. A booklet would contain instructions to reach the next stopping point, where a marked path would lead to the next feature in question. In some cases there would be justification for a specially-prepared wallet (containing a master map), into which the visitor could insert leaflets and guides collected en route. Many of these routes could also form part of conventional trails which would be followed on foot.

11.10   Much of the information regarding motor roads, footpath routes, car parks and other facilities within a Heritage Coast could be combined and presented in the form of a simple inexpensive visitor's map covering the whole of the Heritage Coast. Such a map, which would play a significant role in advising visitors on the location of facilities, should be a priority requirement for each Heritage Coast. It should be attractively produced on the scale of perhaps 1/25,000 or 1/10,000 and obtainable for no more than a shilling or two from information centres and shops in the area as well as in the larger towns nearby. The representative symbol for the area should be included, and the agreed 'house style' adopted for lettering and presentation. Basically a simple folded sheet is envisaged, requiring perhaps the use of two or three colours at most. The main emphasis should be to produce a map that is cheap and readily available but without sacrificing quality of presentation. A map of the kind suggested would show the road system with car parks, picnic places and other focal points, together with the main service facilities and features of interest. All the information shown on the display maps referred to in paragraph 11.5(d) should be included. On its reverse side, the map could contain condensed 'guide book' information, stating clearly where further information and publications could be obtained. Recommended routes and a specimen trail could be included. By means of simple maps of this kind, with concise notes attached, visitors who might not otherwise be inclined to seek out Information Centres could thus be made fully aware of the range of facilities and other information available.

11.11   We envisage that information and interpretative services would be provided jointly by local authorities, private enterprise, and public bodies by means of a co-ordinated policy, with which the English and Welsh Tourist Boards might also be concerned. Close liaison will ensure that the best use is made of the expertise of Nature Conservancy staff, National Trust agents, the managers of private estates, and voluntary organisations representing specialist interests. Such experts should be encouraged to lead the guided walks, and give lectures at the Information Centres and Coastal Centres. They and the wardens are the persons most likely to stimulate and encourage local craftsmen, residents and farmers to welcome and guide visitors.

11.12   The provision of many services and facilities of the types mentioned need not be expensive. A co-ordinated effort will avoid duplication and permit the use of a wider range of resources. Voluntary effort may also be forthcoming. Payment to local authorities towards the cost of information services within National Parks and Areas of Outstanding Natural Beauty and along long distance routes is possible by virtue of s.86(3) of the National Parks and Access to the Countryside Act 1949. By s.12 of the Countryside Act 1968 this is now extended to include study centres and similar facilities within National Parks. The range of information services that may be provided by or with the help of the Countryside Commission is also extended by s.2(8) of the Countryside Act 1968 to apply to places of beauty or interest. By s.5(1) of the 1968 Act payment may also be made to the National Trust and to other persons towards the cost of approved projects. We consider these powers entirely adequate to develop the information and interpretative services which we believe must play an important part in the planning of Heritage Coasts.

CHAPTER 12.
IMPLEMENTATION

12.1 The preceding chapters have outlined the special measures considered necessary to ensure the conservation of Heritage Coasts. In Part 1 the need for these measures has been explained, and in Part 2 areas have been identified where, in the Commission's view, they should be applied. It remains to consider how these policies of conservation and management should be achieved, and who should implement them.

12.2 Three matters may be recalled to the reader's attention:

(1) The need for close liaison between public and private bodies and individuals to ensure that policies of management, access and development do not conflict;

(2) The policy whereby access to, and activity in, a Heritage Coast should not exceed the agreed level for each zone;

(3) The important role that voluntary organisations and members of the public can play.

In implementing the proposals recommended in this study account should be taken of these three factors.

12.3 Circumstances now favour both a new approach towards coastal conservation and the co-operation of public and private agencies on a scale not hitherto envisaged. The Countryside Act 1968 has provided local authorities with a new range of powers and Exchequer grants for conservation and the provision of outdoor recreation facilities. The Town and Country Planning Act 1968 has introduced new elements of planning legislation and placed a new development plan framework at the disposal of some local planning authorities. Much emphasis is now laid on the importance of public participation* and it is now accepted that members of the public have a right to be consulted about, and to take part in, the formulation of local planning policies. As a result of our coastal study, it is now possible to obtain a comprehensive picture of land use and protection along the whole coast of England and Wales, and to identify those undeveloped and scenic stretches in special need of protection. All these factors, therefore, suggest new opportunities.

12.4 Many of the proposals suggested in this study can be implemented under existing legislation and within the development plan structure established by the Town and Country Planning Acts of 1947 and 1962. Indeed, reference to the Summary

* See *People and Planning: a Report of the Committee on Public Participation in Planning*, (M.H.L.G., 1969).

54

of Recommendations (page 62) will show how few of these would require new or amending legislation. Moreover, the realisation is growing that successful rural planning depends on supplementing the controls available under planning legislation by techniques of land management. What is required therefore is a new effort on the part of all concerned, public and private interests alike, to secure the conservation of coastlines of outstanding scenic quality for all time.

12.5   We consider that Heritage Coasts justify special measures beyond those which apply generally in National Parks. Some of the recommended areas fall within National Parks. But special problems arise from the growing and often intense pressures on these narrow strips of coast and threaten serious damage to the landscape and its vegetation cover—far more serious than in most parts of the National Parks. The detailed management necessary within a Heritage Coast would rarely be justified in the much more extensive areas of the National Parks. For these reasons and those mentioned in Chapter 5, a new form of designation is necessary. This would serve both as a legislative framework for the special measures needed to conserve these coasts, and also as a means of recognising their contribution to the national heritage. We recommend that, following consultations with the local planning authorities and other public and private interests, the Countryside Commission should be enabled to designate areas as Heritage Coasts, and to submit proposals for confirmation by the appropriate Minister.*

12.6   The need in Heritage Coasts for special powers has been mentioned in paragraphs 7.12, 7.13, 8.7, 8.20 and 8.29. Availability of additional powers (for example, a Special Development Order for each Heritage Coast) should not lessen in any way the obligation of a local planning authority to secure management and control uses and activities by agreement whenever possible. The stronger measures would be available to a local planning authority as reserve powers, to be used whenever the informal approach proved ineffective and the conservation of the Heritage Coast might be imperilled. (Appropriate measures which might be covered by agreement are suggested in paragraph 12.19.)

12.7   Throughout this study it has been assumed that the local planning authority will continue, and on an increasing scale, to play the major role in the day to day control of land use and development in a Heritage Coast. Because of the range of measures needed it is considered that no other agency would be as effective in promoting the conservation of these areas. But the measures suggested involve a degree of positive land management not contemplated in the Town and Country Planning Acts. Skills and expertise in the fields of conservation and estate management are required.

12.8   We recommend that for each Heritage Coast it should be a requirement that the local planning authority (or, if more than one is involved, the local planning authorities acting in agreement) should delegate to a special committee the planning and management functions necessary to secure the objectives set out in paragraph 6.4. Each scheme of delegation should require Ministerial approval. The committee should include (as in the National Parks) nationally appointed members representing, for example, landowners, agriculturalists, and those concerned with nature conservation and recreational interests. The arrangement must be flexible to meet both the particular needs of the area and also the structure and organisation of the local

---

* In England, the Minister of Housing and Local Government, or, in Wales, the Secretary of State for Wales.

planning authority. In some cases the committee might be a sub-committee of the Planning Committee, or alternatively a Countryside Committee or a special Heritage Coast Committee. Whichever course is adopted the Minister must ensure that the committee can draw on adequate financial and staff resources to carry out its duties effectively.

12.9 The responsibilities of the committee would be the formulation of the overall policy for the Heritage Coast, including the co-ordination of management policies covering the control of development, the provision of recreational facilities, vehicular access, footpaths and bridleways, access to open country, access from the sea, landscape improvement and information and interpretative services. The committee would also determine acceptable levels of use and define the management zones in the manner described in Chapter 6.

12.10 We consider it essential that a suitably trained Conservation Officer should be appointed for each Heritage Coast to advise the committee on the management issues. We see this as a local government appointment, probably to the staff of the County Planning Officer, but with an appropriate degree of delegated authority. In paragraph 12.26 we make recommendations as to grant aid for the salary of the Conservation Officer. The remuneration should be related to the size of the area and the pressures to which it is subjected. Where the Heritage Coast is small in extent, a full-time specialist appointment may not be justified. In such cases it would seem likely that the responsibilities of this post would also include other duties exercised by the local authority, and a suitable apportionment of time and salary would need to be agreed.

12.11 In addition to a Conservation Officer, whose work would be divided between the local planning authority's administrative headquarters and the coast itself, staff, including wardens, would be required to carry out the practical tasks of management. We see a clear distinction between the functions of a warden and those of a Conservation Officer. The former would construct and maintain the necessary paths and signs, and guide and assist visitors; a Conservation Officer would be concerned more with the co-ordination of policies and the selection of appropriate management techniques, for example to choose appropriate measures to control visitors (see paragraph 6.13), to determine the rotation of paths, to monitor the responses to the range of facilities provided (and the effects of different levels of use on the environment) and make adjustments accordingly (as described in paragraph 6.11), and to determine at which point the persuasive measures of traffic management described in paragraph 8.5 become ineffective, and stronger controls are needed. These call for specialist skills and experience backed by appropriate professional or academic qualifications. The Conservation Officer must also be able to instruct the wardens as necessary.

12.12 Many of the practical tasks of construction and maintenance, and preparing facilities for visitors, would be carried out during the period from late autumn to early spring when there are few visitors. Many of these tasks could be undertaken by local authority staff engaged on other duties at busier times. The Conservation Officer should be able to train such staff to function as wardens.

12.13 The management policies of a Heritage Coast would be implemented, in practice, by a combination of various methods. The management policies of public bodies such as the Forestry Commission and the Nature Conservancy would, very often, be in line with the overall objectives. Sometimes it would be necessary to initiate management agreements with private landowners, farmers, clubs and societies and field study centres, on the lines suggested in paragraph 12.19. In some places

10. *Purbeck:* Ballard Down, north of Swanage.

11. *Chesil Beach:* looking north-west towards Abbotsbury.

12. *Lyme Bay:* Golden Cap and Thorncombe Beacon, near Chideock.
13. *Scabbacombe Head:* Mill Bay Cove and Newfoundland Cove, south of Kingswear.
14. *South Devon:* Gammon Head and Prawle Point.

15. *Rame Head:* with
Whitesand Bay
beyond.

16. *Polperro-Gribbin
Head:* Parson's Cove
and West Coombe,
near Lansallos.

17. *Mevagissey-Zone Point:* Veryan Bay and Dodman Point.

18. *Lizard:* Lizard Point, looking towards Kynance Cove.

19. *West Penwith:*
Porthgwarra and
Hella Point, south of
Land's End, looking
towards Porthcurno.

20. *Isles of Scilly:*
Porth Hellick, St.
Mary's, with Hugh
Town beyond.

21. *Tintagel-Widemouth:* Tintagel Head, looking towards Willapark.

22. *Hartland:* Vicarage Cliff, Morwenstow, and Higher Sharpnose Point.

23. *Exmoor:* Great Hangman and Holdstone Down, east of Combe Martin, looking to Foreland Point.

24. *South Glamorgan:* Traeth Bach, Dunraven, looking towards Nash Point.

25. *Gower:* Burry Holms and Broughton Bay, with Whitford Burrows and the Loughor estuary beyond.

access agreements would be necessary. All of the above would be appropriate within remote zones and in those parts of transitional zones likely to attract smaller numbers where the land would remain in private ownership subject to access safeguards.

12.14   In some key areas public or quasi-public ownership would be the only effective long term safeguard. The National Trust, for instance, may be prepared to purchase specific stretches of coastal land, or to consolidate existing Trust ownerships, or to enter into covenants with landowners. Ownership by the local authority would be appropriate where other measures are ineffective or unduly expensive—for example, for the removal of disfigurements, the initiation of a land reclamation scheme, or the provision of recreational facilities of a standard and in places where private enterprise is unlikely or unwilling to provide them. Acquisition is only one way in which management can be facilitated. It should not be used to secure the control of development which could be achieved either by statutory planning control or by voluntary co-operation and goodwill.

12.15   There must be regular and close consultation with public bodies for drawing up codes of management and co-ordinating policies. Organisations including the Ministry of Defence, the Post Office, the Forestry Commission, the Nature Conservancy, Trinity House, electricity, gas and water undertakings may be involved in managing parts of a Heritage Coast or in carrying out building or engineering work within it. The Conservation Officer should convene meetings with public bodies to draw up and keep under review codes of management to ensure that policies applying to land held or managed by these bodies in a Heritage Coast do not conflict either with each other or with the general management objectives.

12.16   Matters for discussion would include, for example, access to Ministry of Defence land and to nature reserves, measures to secure traffic management, agreement for the use of forest roads as part of a scenic route (paragraph 8.13), the rationalisation of existing recreational facilities (paragraph 9.7), improvement of the landscape by re-routeing overhead lines or improving views (paragraphs 10.3 and 10.4), and the co-ordination of information services (paragraph 11.11).

12.17   There must be effective liaison with the National Trust, private landowners, farmers, voluntary organisations, amenity societies and other interested parties in order to provide the link between the committee referred to in paragraph 12.8 and those who live or work in the area or have a special interest in it. At meetings with individuals or organisations, or groups of organisations, the Conservation Officer would aim, firstly, to explain the policies of his committee so that representatives are kept fully informed and are able to see how they can help, secondly to establish codes of management for agricultural and other operations, and thirdly, to foster an atmosphere of goodwill which may make it possible for the local planning authority to enter into agreements for specific management objectives.

12.18   The agenda of these meetings and discussions might include such items as (1) measures to minimise conflict between walkers and agricultural interests (see paragraph 8.17), (2) the provision of access to open country (see paragraph 8.28), (3) the co-ordination of interpretative services (see paragraph 11.11) and tourist facilities (for example to confine, where practicable, commercial holiday motor boat trips which involve landings to those parts of the coast where least damage is likely to be caused), (4) the siting and design of signs and notices (paragraph 10.6), (5) measures to improve the appearance of the landscape, and (6) codes of practice to cover aspects such as tree felling, the use of chemical sprays, the siting of new buildings, the removal of hedges and the ploughing of open land on cliff tops.

*E

12.19   As a result of the goodwill that it is hoped would be created by this policy of consultation, a local planning authority should be able more readily to obtain the agreement of private landowners, farmers and others for carrying out specific management objectives. In our main report* we advocate the use of management agreements, which could include positive as well as restrictive covenants. Agreements made between a local planning authority and a private individual or body should be widely used in a Heritage Coast to secure positive management, without the need for compulsory acquisition. In return for financial help, an individual might agree (1) to improve or maintain access by clearing paths, installing cattle grids, or providing car turning spaces, (2) to provide temporary parking facilities at agreed locations but not elsewhere on his land (see paragraph 8.12), (3) to allow camping only on certain parts of his land, or not to permit any camping at all (see paragraph 9.10), (4) to erect signs or display maps to guide visitors, or make certain facilities available to the general public (see paragraph 9.7), (5) to improve the appearance of the landscape or open up views by removing or repairing unsightly buildings or structures (see paragraph 10.4), or (6) to clear litter periodically from the less accessible parts of the coast (see paragraph 10.5). A local planning authority should also be able to make agreements with recreational clubs and societies, to regulate the number of users of a specific facility or site, or with field study centres to control the numbers of students at any one time carrying out an intensive study of ecological habitats which are particularly vulnerable to disturbance.

12.20   Usually a management agreement is likely to prove the cheapest and most effective way of securing objectives which cannot be obtained by planning control, and would be much more acceptable than compulsory acquisition. In addition to receiving financial consideration for any services rendered and for any detriment to his interest, a landowner entering into an agreement would frequently benefit by the presence of wardens, who would guide visitors and minimise disturbance. Where repeated efforts to secure agreement do not materialise, a local planning authority should consider, as a last resort, the use of stronger measures of the kinds referred to in paragraph 12.6.

12.21   Unfortunately, unlike restrictive agreements made under Section 37 of the Town and Country Planning Act, 1962, management agreements which contain positive covenants present legal difficulties. More comprehensive legislation, to enable public authorities to make positive agreements with private owners and, where necessary, to bind successors in title, is long overdue and should be introduced at the earliest opportunity.

12.22   In order to increase public support for measures to conserve the unspoilt scenic stretches of coast, and to co-ordinate voluntary effort and apply it to the tasks of management, the Conservation Officer should organise, and have the financial resources available to him to support, work by volunteers sympathetic to the management objectives who would be able to help in promoting the conservation of the area and in carrying out specific tasks under his guidance and that of his staff. Experience suggests that many offers of voluntary assistance are not taken up because of the lack of an organiser who would not only harness and apply this enthusiasm but who would also arrange insurance cover (if necessary), supply equipment, and arrange transport and possibly accommodation. Although most volunteers would be local persons, visitors from further afield who are able to help should be encouraged to take part in

* *The Planning of the Coastline*, para. 218.

58

these activities. Examples of work that might be undertaken include footpath clearance, the erection and maintenance of signs, display boards, and panoramas, the manning of information posts, litter clearance and voluntary warden duties.

12.23 As mentioned in Chapter 11, considerable importance is attached to the building up of a body of interested persons who know the coast and countryside. Public interest in a particular stretch of Heritage Coast is likely to be enhanced by giving an 'identity' to the coast in question (perhaps by adopting the 'house symbol' as a badge to be worn on armbands, and so forth) and by arranging lectures on local history, wild life and other aspects of the Heritage Coast.

12.24 The overall policy outlined in Chapter 6 must be embodied in some official plan. In the long term the new style development plans, introduced by the Town and Country Planning Act 1968, will apply to the whole of England and Wales. The Heritage Coast boundary and the main policy objectives would then be indicated on the structure plan. A district plan for the Heritage Coast would contain, for each of the management zones, a policy statement setting out the codes on which management and development should be based, with guide lines for the management of land held by public bodies. Superimposed on this would be detailed management proposals for specific sites within the Heritage Coast.

12.25 It will however be many years before all local planning authorities are authorised to use the new development plan system, and few Heritage Coasts are likely to be within areas included for early approval. We consider that, within two years of designation, a local plan for each Heritage Coast should be submitted for approval by the Minister. The plans should adequately reflect the national significance of these areas. We anticipate that the Minister would seek our comments before such a plan is approved. It is suggested that the Minister should exercise his powers under Schedule 10 to the Town and Country Planning Act 1968 and require local planning authorities whose areas include Heritage Coasts to produce amendments to current development plans on the lines outlined in Sections 6 to 9 of the 1968 Act regarding the preparation of local plans.

12.26 To ensure that the policies implemented at local level are adequate to safeguard Heritage Coasts, we consider that some national stake in these coasts is essential. We propose, therefore, that within each Heritage Coast, in addition to the existing range of Exchequer grants available, the salary of a full-time Conservation Officer (or a negotiated proportion of a part-time officer), and the administrative costs of Heritage Coast management, should be eligible for 90% Exchequer grant.* We recommend this special arrangement for several reasons. Firstly, most of the Heritage Coasts will fall within areas of low rateable value where, because of the restrictions on new development, the opportunities for enhancement will be slight. Secondly, many of the proposals we recommend cover new fields of planning and management in which there is little activity at local planning authority level, and we think that special encouragement will be needed to ensure that schemes are implemented. In the absence of such an arrangement it is likely that the burden of staff and administrative costs will be so great as to prevent the introduction of the planning and management measures we describe in this report, and which we consider so essential. Above all we are anxious to ensure that the status of the Conservation Officer, the

* The annual cost to the Exchequer of these measures would be unlikely to exceed £200,000 if all our proposed areas were designated. This assumes grant payment for the salary and special equipment of a Conservation Officer but excludes other items such as picnic sites within a Heritage Coast, for which grant aid may also be payable.

responsible committee, and the ancillary staff and resources at their disposal, suitably reflects the national objectives which underlie the conservation of these coasts. We believe that Heritage Coasts represent the best examples of natural coastal scenery in England and Wales, and that the pressures on them are likely to be greater than in many National Parks. Nothing less than this degree of public involvement will secure the conservation of these coasts for all time.

Part Four

Summary of Recommendations

SUMMARY OF RECOMMENDATIONS

Numbers refer to paragraphs in the report.

Part 1.   The Need for a Policy (Chapters 1, 2 and 3)

3.1   The finest stretches of coast justify a special claim for protection and should be nationally identified and managed comprehensively; stronger measures than elsewhere should apply.

Part 2.   National Review of Scenic Coastlines (Chapters 4 and 5)

5.5   Those areas selected by the Commission as described in Part 2 should be designated, nationally, as Heritage Coasts.

Part 3.   Heritage Coasts

Chapter 6.   Overall policy and principles of management

6.1   The planning objectives for a Heritage Coast should be to conserve the natural coastal scenery and to facilitate and enhance its enjoyment by the public.

6.3   The aim should be to facilitate the enjoyment of the natural qualities of the coast without introducing artificial attractions. With careful management some parts of Heritage Coasts should continue to accommodate large numbers of visitors.

6.4   The following management principles should be applied:

(1) The scale of recreational activities, and of facilities that are provided, should be related to an optimum level of use.

(2) For management purposes the coast should be divided into zones, each with an agreed level of use.

(3) Rigorous control should be exercised over unsuitable development.

(4) Access should be regulated to avoid concentrations of persons and vehicles at points liable to damage.

(5) Schemes should be initiated to enhance the appearance of the landscape.

(6) The opportunities for different forms of recreation should be emphasised, so as to permit a greater use of existing resources without the need for new facilities.

(7) Information services and interpretative techniques should be introduced to encourage greater interest in the environment.

Determination of intensity of use

6.9   Further research should be initiated into the carrying capacity of resources and their ability to withstand public pressure.

6.11   Until capacity can be determined quantitatively, a flexible approach should be adopted which enables the effects on the environment of different levels of use to be assessed, and policies to be adjusted accordingly.

Determination of management zones based on different intensities of use

6.17   The management policy for an *intensive zone* should ensure that the facilities are adequate and are designed to minimise any adverse effects on the landscape.

6.19   The management policy for a *remote zone* should keep it free from vehicles.

6.21   The management policy for a *transitional zone* should group facilities in selected places, and retain the stretches between them as unspoilt coast with a reasonable degree of road access.

6.22   An initial survey and appraisal of existing and potential resources, such as that shown in Appendix 2 to this report, should form the basis for determining the principal management zones.

Chapter 7.   Land use and development

Non-recreational land uses

7.3   Only those existing uses which do not justify removal, or are the subject of outstanding consents, as well as agriculture, forestry, nature conservation and coastal navigation or safety, should be regarded as acceptable within a Heritage Coast.

7.4   Special care should be given to the siting and design of all new buildings and works which are absolutely necessary in connection with acceptable uses.

7.5   Other uses should be regarded as unacceptable unless their national importance would justify a site within a Heritage Coast.

7.12   A Special Development Order should be introduced to make the provisions of article 4 of the Town and Country Planning General Development Order 1963 apply to each Heritage Coast, having effect as though a direction were in force to control permitted development.

7.13   The Special Development Order should define each Heritage Coast as an area of special control of advertisements.

Recreational land uses

7.14   Only where a recreational activity is acceptable should supporting facilities be accepted or provided.

7.15   Only those activities should be encouraged which (1) have a minimum impact on the landscape, (2) rely on the natural resources of the coast, and (3) do not conflict with other activities.

7.16   The determination of acceptable activities should be based on the management zones, on the lines of the Chart of Acceptable Activities included in Chapter 7.

7.17 Local planning authorities should adopt similar measures, as part of the development plan, to determine which recreational uses are acceptable. Where development control alone would be insufficient to prevent inappropriate activities, measures such as byelaws or speed limits should be introduced.

Chapter 8. Access

Vehicular access

8.3 The management policy should have the following aims:

(a) and (b) Road and parking space capacities should be related to the acceptable level of use.

(c) Through traffic should be discouraged.

(d) Concentrations of parked vehicles should be avoided at points where damage is likely to occur, and vehicles should not be permitted on beaches.

(e) Access within remote zones should be confined to essential service vehicles and those requiring access to isolated buildings.

(f) No improvement or construction of roads should be carried out within remote zones which would lead to increased penetration by non-essential vehicles.

(g) In transitional zones, vehicles should be kept away from areas vulnerable to damage.

(h) Where conflicts arise from existing pressures in transitional zones, short lengths of road should be stopped up and the car parks re-sited to provide foot access.

(i) Measures should be introduced to provide a one-way system, or to restrict the type of vehicles using certain roads.

(j) Within intensive zones the principal features and beaches should be provided with convenient road access.

8.5 Every effort should be made to channel motorists along certain suitable routes by means of signs which encourage and invite entry; prohibition of access to less suitable routes should be adopted only as a last resort.

8.6 Local planning authorities should seek the agreement of farmers and landowners in controlling the use by the public of private roads and motorable tracks.

8.7 The provisions of s.32 of the Countryside Act 1968 with regard to traffic regulation should be extended to apply to all land in Heritage Coasts.

Parking facilities

8.10 Parking places should be provided at settlements, beaches, roadheads, viewpoints, picnic areas, features of special interest, and on scenic roads.

8.12 Local planning authorities should be encouraged to make informal agreements with landowners to provide and control temporary car parking facilities at peak periods.

Scenic roads

8.13 Where special opportunities occur, scenic roads should be provided in transitional zones.

8.14 The design of scenic roads should discourage through traffic.

Footpaths

8.16 The management policy should have the following aims:

(a) The footpaths should allow convenient walks between focal points, as an alternative to roads.

(b) Paths should avoid places where walkers may damage the habitat, and should lead them to specific areas where access is to be encouraged.

(c) Paths should be planned so that they avoid roads, or, if this is impossible, follow them for only short distances.

(d) Where a path ends abruptly at a road, a new link path should be provided.

(e) A path should be provided along the cliff or shore of each Heritage Coast.

(f) Alternative return routes, for use with the cliff or shore path, should also be provided.

(g) Paths should be provided along the lines of the major linear topographical features of the area.

(h) Signposting should be unobtrusive, yet adequate, and in remote zones confined to simple waymarking. The use of signs smaller than those prescribed in the Traffic Signs Regulations 1964 should be encouraged.

(i) The existence of routes and alternatives, with times, distances and other details, should be adequately publicised so that a visitor may pick the route which suits his special needs.

8.17 Local planning authorities should maintain close liaison with highway authorities, Government Departments and magistrates' courts to co-ordinate footpath policy. Similar liaison with landowners and farmers should take place to minimise any conflicts with agricultural interests.

8.18 Every opportunity should be taken to encourage local voluntary organisations to play a part in signposting and marking paths.

8.20 The national significance of a path along the cliffs and shore of Heritage Coasts should be recognised by the payment of Exchequer grant as if the path were part of a long distance route established under s.51 of the National Parks and Access to the Countryside Act 1949.

Bridleways

8.23 The bridleway network should consist of the trunk portions of the footpath network, especially those which link settlements, beaches, stables, picnic places and viewpoints.

8.24 The consequences of providing bridleway access to remote zones, and to any areas likely to suffer damage, should be carefully considered.

8.25 Local authorities should take advantage of the special review of 'roads used as public paths' (Schedule 3, Part III, of the Countryside Act 1968) to re-classify roads as bridleways to provide necessary links where vehicular use cannot be justified. The desirability of byelaws which restrict the use of pedal cycles on specified bridleways outside villages should be reconsidered.

8.26 Close consultation with landowners, farmers and local interests should be effected to co-ordinate policies and to secure management techniques.

Access to open country

8.27 Wherever possible, uncultivated land immediately adjoining the shore, and having special recreational value, should be available for public access.

8.28 Local planning authorities should co-ordinate the management policies of landowners and public bodies which relate to open country.

8.29 The provisions of s.14 of the Countryside Act 1968 regarding the ploughing of moorland or heath should be extended to all such land in Heritage Coasts.

Access from the sea

8.31 Local planning authorities should ensure that the degree of access possible from the sea is consistent with the management policy for the appropriate zone.

8.32 Direct access from the sea should be confined to specified points, and byelaws should be introduced.

Chapter 9. Provision of recreational facilities

9.2 Provision should be made only for those facilities which are sufficient to allow full enjoyment of the scenery and activities based on it. Most facilities should be channelled into intensive zones. The adaptation of existing buildings or structures, as a means of providing facilities, should be encouraged.

9.3 Facilities should be grouped at principal focal points and not scattered throughout the Heritage Coast.

9.5 The determination of what are appropriate facilities should be based partly on the management zone and partly on the nature of the focal point at which they are grouped, in a manner similar to that shown on the Facility Provision Chart included in Chapter 9.

9.7 Local planning authorities should adopt, as part of the development plan, measures such as those described in this chapter to determine which facilities are appropriate. Local planning authorities should also decide if any existing facilities should be re-sited and grouped at focal points. Arrangements should be encouraged to permit public use of facilities hitherto reserved for specified groups of users.

9.8 New buildings providing accommodation for visitors in a Heritage Coast should be regarded as unacceptable outside existing settlements.

9.9 The conversion or adaptation of existing buildings to provide accommodation for visitors should be regarded as acceptable if there is no detrimental effect on the landscape, and if the building is attractive and would otherwise fall into disrepair.

9.10 Motorised camping should normally be regarded as unacceptable in most parts of a Heritage Coast.

Chapter 10. Landscape improvement

10.2 Local authorities should be required, as soon as boundaries have been agreed, to carry out a review of each Heritage Coast to decide on which sites improvement schemes should be carried out.

10.3 There should be co-ordination of management policies of statutory undertakers and public agencies, for example to improve the siting of overhead lines, and to create and improve views by the removal of obsolete buildings and advertisements.

10.4 There should be consultation with private owners to minimise any features which disfigure the landscape or prevent full enjoyment of the scenery.

10.5 Local planning authorities should be encouraged to initiate agreements with farmers and others for the clearance of litter from remote zones, and from the less accessible parts of transitional zones.

10.6 The number and size of all signs and notices should be kept to a minimum.

10.7 Special care should be exercised in the design of new notices by the use of modern lettering and, if necessary, pictorial symbols.

Chapter 11. Information and Interpretation

11.3 The policy should (1) emphasise the recreational opportunities, (2) promote those which are acceptable, (3) advise about parking spaces and facilities at peak periods, (4) explain the reasons for the methods of management, and (5) stress any local dangers.

11.5 The methods and techniques employed should include the provision of Information Centres and Posts, Coastal Centres, the Warden Service, display maps, guided walks, trails, panoramic viewfinders, observation hides, lectures, film shows, exhibitions, displays, demonstrations of local crafts and activities, and a visitor's map.

11.6 The local identity of a Heritage Coast should be emphasised by means of a symbol and a 'house style' for use on signs and publications.

11.8 A variety of self-guided trails should be devised, covering a more extensive range of features than the conventional nature trail.

11.9 Special self-guided motor trails should be devised, particularly for use on scenic roads.

11.10 A simple inexpensive visitor's map should be produced for distribution as a priority requirement for each Heritage Coast. It should also give concise information on facilities available.

11.11 Information and interpretative services within a Heritage Coast should be produced jointly by local authorities, private enterprise, public bodies and voluntary organisations as part of a co-ordinated policy.

Chapter 12. Implementation

12.5 After consultations, the Countryside Commission should be enabled to designate a Heritage Coast and submit proposals to the Minister (or Secretary of State for Wales) for confirmation.

12.6 The availability of special powers and stronger measures than would apply outside Heritage Coasts should not lessen the local planning authority's obligation to secure management and control uses by agreement whenever possible.

12.7 Local planning authorities should continue to play the major role in controlling land use and development within Heritage Coasts.

12.8   For each Heritage Coast a local planning authority should be required to delegate the planning and management functions to a special committee, which should include nationally appointed representatives. Each scheme of delegation should require Ministerial approval, and the Minister should ensure that the committee is able to draw on adequate financial and staff resources.

12.9   The committee should formulate the overall policy for the Heritage Coast and co-ordinate policies of development, access, landscape improvement, information and the provision of recreational facilities.

12.10   A suitably-trained Conservation Officer should be appointed for each Heritage Coast to advise the committee on the management issues involved. This should be a local government appointment. Where a full-time appointment for this purpose is not justified, the post should include other responsibilities, and an apportionment of time and salary should be made accordingly.

12.11   The Conservation Officer should be provided with staff, including wardens, to carry out the practical tasks of management.

12.12   The Conservation Officer should be able to train certain other local authority staff to function as wardens.

12.13   Management policies should be implemented by a combination of methods including management agreements, access agreements and the co-ordination of existing policies operated by public bodies.

12.14   Complete public ownership should be considered in key areas if other methods are ineffective or too expensive; it should not be used to secure the control of development for which other measures are available.

12.15   The Conservation Officer should convene meetings with public bodies involved in managing parts of a Heritage Coast in order to co-ordinate policies and to draw up codes of management.

12.17   Liaison with the National Trust, private landowners, farmers, voluntary organisations and others should be effected in order to provide the link between the committee and those who live or work in the area. At meetings with such persons and bodies the Conservation Officer should aim to establish codes of management for agricultural and other operations, and seek to foster the atmosphere of goodwill that would facilitate the making of management agreements.

12.19   Management agreements should be widely used to secure positive management without the need for compulsory acquisition.

12.21   Legislation should be introduced at the earliest opportunity to enable public authorities to make positive agreements with private owners, which are enforceable and which bind successors in title.

12.22   The Conservation Officer should have the financial resources to support the efforts of volunteers and should be able to harness their enthusiasm in tasks of management. He should organise the carrying out of specific tasks and provide the necessary administrative services.

12.24   In the long term the plan for a Heritage Coast should be embodied in the new style development plan introduced by the Town and Country Planning Act 1968. The main policy objectives should be indicated on the structure plan. A district plan should contain a policy statement for each of the management zones, and set out the codes on which management and development should be based. These should include guide lines for the management of land held by public bodies.

12.25   Until the 1968 Act is fully operative throughout England and Wales, the powers under Schedule 10 to the 1968 Act should be used. Local planning authorities should be required to make amendments to current development plans to safeguard Heritage Coasts. Within two years of designation a local plan for each Heritage Coast should be submitted for approval by the Minister. The plan should adequately reflect the coast's national significance.

12.26   The salary of a Conservation Officer and the administrative costs of management within each Heritage Coast should be eligible for special 90% Exchequer grant.

APPENDIX 1.
Descriptions and main features of proposed Heritage Coasts*

1. NORTH NORTHUMBERLAND

(a) NT 979576—NU 006528; (b) NU 038475—NU 258202. Total length 50·6 miles, of which 2·5 miles are substantially developed.

This stretch of coast runs from the Border to Craster; it is divided into two parts by the built-up area around Berwick. That to the north of the town is formed of Carboniferous rocks which can be examined in the cliffs. That to the south is at first low and rocky, but soon widens out to form the extensive sand flats which reach out to Holy Island. Ross Links separate these flats from Budle Bay, another shallow inlet. Immediately south of this bay the Whin Sill is the dominating feature; Bamburgh Castle is built on it, and the Farne Islands are severed portions of it.

Low, rocky, headlands and small havens and bays with fine beaches reach as far as (Dunstanburgh) Castle Point where the Whin Sill, resting on limestone, is seen to great advantage. On the south side of the Point the Whin outcrops on the foreshore.

The Farne Islands are owned by the National Trust; the tidal flats between Holy Island, Ross Links, and Budle Bay are included within the Lindisfarne National Nature Reserve, the chief interest of which is ornithological.

This coast has many historical and ecclesiastical associations. The names of Saint Cuthbert, Saint Aidan, and of King Ida (the Flamebearer) of Bamburgh Castle are as familiar as that of the Abbey of Lindisfarne and the much later exploit of Grace Darling.

2. NORTH YORKSHIRE

NZ 715202—SE 035912. Length 33·1 miles, of which 2·5 miles are substantially developed.

The North Yorkshire coastal area is roughly coincident with the seaward edge of the North York Moors National Park. It runs from Skinningrove to Scalby Ness.

It is a cliffed coast, formed of Jurassic rocks, and landslips have locally given rise to undercliffs. Boulby, where alum quarrying has taken place, was once the highest cliff in England. There are three well known bays—Runswick, Sandsend and Robin Hood's Bay—and many smaller openings, or *wykes*, as at Staithes and Hayburn. The

---

* The areas, which are shown on Map 2, are described here in clockwise fashion, working south from Northumberland.

70

mouth of the Esk at Whitby is noteworthy. Beaches are not extensive; the largest is at Sandsend. Boulder clay covers much of the solid rocks, and the larger bays and many of the smaller inlets are cut in it.

Hayburn Wyke is a reserve of botanical significance belonging to the Yorkshire Naturalists' Trust. The long distance footpath, the Cleveland Way, follows the cliff edge. There are several fishing settlements, including Whitby, which is also distinguished by the ruins of the Abbey on the cliff top. There are also associations with Captain Cook.

3. FLAMBOROUGH HEAD

TA 152756—TA 203687. Length 11·8 miles, of which 0·25 miles are substantially developed.

Flamborough is a chalk headland covered with a thick mass of boulder clay which slopes at a fairly steep angle above vertical chalk cliffs in which several caves and stacks have been carved. The chalk cliffs on the north side of the headland are picturesque, and are continuous with those at Bempton. It is perhaps the finest line of chalk cliffs in the country. At Speeton in the north and at Sewerby in the south the chalk gives way to boulder clay, and at Speeton there are landslips.

Flamborough Head is one of the major headlands on the east coast, and is much visited. The Danes' Dyke crosses from north to south and there is much to interest geologists and ornithologists. There is also a prominent lighthouse.

4. SPURN HEAD

TA 416152—TA 421153. Length 7·2 miles.

Spurn Head is a sand spit which has formed across the mouth of the Humber, largely as a result of the rapid erosion of the boulder clay cliffs of Holderness. On its western side are extensive mudflats. The evolution of the spit is of interest to physiographers, and for that reason and also because of the great number of sea birds associated with it, it has been designated a Site of Special Scientific Interest. Part of it is a reserve managed by the Yorkshire Naturalists' Trust.

There is a road along the spit leading to a lighthouse and coastguard station. Owing to the movements of the spit, the lighthouse has had to be rebuilt several times.

5. NORTH NORFOLK

TF 693438—TG 093442. Length 39·3 miles.

The marshland coast between Holme next the Sea and Weybourne is the finest example of its kind in these islands, and probably in Europe. The old coast can be traced as a gentle slope up from the marshes, and the present coast consists of wide sandy beaches, usually with some shingle, broken by natural marsh channels, used, mainly in the past, for access to small ports. Where the marsh has been reclaimed there is, apart from a few unfilled channels, no break between the outer beach and the old cliff. In other places barrier beaches and shingle spits have grown in such a way as to enclose large areas of undrained marsh. Scolt Head Island and Lodge Marsh (Wells) are barriers; Blakeney Point, along which shingle predominates, is a spit. All these features are faced on the seaward side by wide sandy beaches which, when dry, afford abundant supplies of sand to build the numerous lines of dunes. At Holkham at low water of spring tides nearly two miles of sand are exposed.

The beaches, shingle and sand, the dunes and the marshes are in a constant state of evolution, and changes of some magnitude are not infrequent. For this reason the coast is of major interest to all students of shoreline processes. The variety of plants in all three habitats, beaches, dunes and marshes, and the effects they have on the accumulation of sand and mud make this coast a supreme laboratory for the study of ecology and its interrelationships with physiography.

Blakeney Point and Scolt Head Island and their terneries are known to ornithologists throughout the world. Because of its outstanding importance, the coast is almost wholly protected. The National Trust, the Nature Conservancy and three local bodies own and administer parts of it. There are National Nature Reserves at Holkham and Scolt Head, and the area is designated as an Area of Outstanding Natural Beauty. Recently a joint committee representing the bodies mentioned, and also the Norfolk County Council, has been formed to co-ordinate policy.

Inland lies a string of attractive villages and harbours, including Brancaster, Overy Staithe and Blakeney. There are several windmills in the vicinity. The coast is associated with the Norwich School of painters, notably Crome. The Peddars Way, a prehistoric track, meets the coast in the west, at Holme next the Sea.

6. SUFFOLK

TM 537850—TM 332376. Length 34·1 miles, of which 4·0 miles are substantially developed.

Although within a relatively short distance of London, the coast of Suffolk between Benacre and the mouth of the Deben is largely unspoiled. It consists of lines of low cliffs of soft and incoherent sands alternating with what were at one time small inlets or river mouths, e.g. Covehithe Broad, Easton Bavents, and Minsmere. The travel of beach material is to the south, and in past times the Blyth and the former Dunwich river often shared a common mouth which shifted with the movement of the shingle. The Blyth is now held in position at Walberswick, but the Alde (Ore) is deflected ten or eleven miles to the south by the great shingle foreland of Orford Ness.

There is a close relationship between the rise and fall of small ports and the changes of this coast. At Dunwich erosion has been serious over the centuries, and the medieval town has disappeared. Orford's former importance, indicated in part by its castle, has fluctuated and decayed with the growth of the shingle spit. Southwold originally stood on an island; the main road into the town crosses Buss Creek, which in earlier times was an open estuary.

The upper parts of the Deben, Alde, and Blyth are very attractive, and several of the small towns and villages on or near the coast are notable for the beauty of their churches. Minsmere and Havergate Island (inside Orford Ness) are nature reserves managed by the R.S.P.B. Westleton Heath and Orfordness-Havergate are National Nature Reserves.

In recent years Aldeburgh has become well-known as a result of its annual music festival; the poet Crabbe also lived here. There are also associations with Benjamin Britten and the painters Charles Keene and John Sell Cotman.

7. SOUTH FORELAND

TR 379464—TR 340422. Length 3·9 miles.

This is only a short stretch of coast, but is well known because it is the part so commonly referred to as the White Cliffs of Dover. These are nearly vertical cliffs of chalk. The gap formed by St. Margaret's Bay is a good feature, and the whole stretch is popular with visitors. There is an interesting lighthouse.

72

26. *South Pembrokeshire:* Manorbier Bay, with Old Castle Head beyond.

27. *Marloes and Dale:* St. Ann's Head, at the entrance to Milford Haven.

28. *North-West Pembrokeshire:* Porth Maen-melyn, near Strumble Head.

29. *St. Dogmaels:* Pwllygranant and Pen-yr-afr, looking towards Cemaes Head.

30. *Lleyn:* Bardsey
Island, off the south-
western end of the
peninsula.

31. *Holyhead
Mountain:* South
Stack, with Holyhead
Mountain behind.

32. *North Anglesey:*
Porthybribys and
Carmel Head, near
Llanfairynghornwy.

33. *Great Orme:* The northern cliffs and Marine Drive.

Below:
34. *St. Bees Head:* The cliffs of North Head, with Saltom Bay beyond.

8. SEVEN SISTERS AND BEACHY HEAD

TV 600967—TV 489981. Length 8·3 miles, of which 0·1 miles are substa
developed.

The North Downs reach the sea at the South Foreland; the South Downs
it here at Beachy Head. The views from both Beachy Head and Seaford He
magnificent. The Seven Sisters, partly owned by the National Trust, are relics of the
chalk surface separated by valleys which are now dry. The valley mouths are high
above sea level. Today the sea is eroding the cliffs and in this way is, at any rate in
part, responsible for the height of the valley mouths. This piece of coast also includes
the Cuckmere valley with its fine meanders. The chalk grassland on the cliff tops is
noteworthy.

The South Downs Long Distance Bridleway follows the cliffs, and provides access
to ancient earthworks, tumuli, and a hill fort. There are good views throughout.

9. ISLE OF WIGHT

SZ 541765—SZ 318865. Length 20·4 miles, of which 0·5 miles are substantially
developed.

The chalk of Hampshire passes under the Solent to reappear as the fine ridge running
from the Needles to Culver Cliff. The beds beneath the chalk—the Lower Greensand
and the Wealden Clays—are, like the chalk, folded, and in consequence the line of
cliffs from the Needles to St. Catherine's Point is made up of several types of rock.
North of the Needles the beds above the chalk reach the coast, and the Tertiary beds
in Alum Bay include the well-known coloured sands. The Needles, probably one of
the best-known coastal features in Britain, are stacks of chalk which owe their preservation largely to the hardness produced by intense folding. At St. Catherine's Point,
where there is a lighthouse, the Greensand has slipped and formed an undercliff in
front of the chalk above.

There are Sites of Special Scientific Interest, and the high chalk cliffs at and near
Tennyson Down (named after the poet, who lived at Farringford House) are noteworthy. Several stretches are owned by the National Trust.

10. PURBECK

(a) SZ 042825—SZ 035805; (b) SZ 037784—SY 760814. Total length 24·6 miles, of
which 0·1 miles are substantially developed.

The Isle of Purbeck closely resembles the Isle of Wight in structure, and at an
earlier time the chalk ridge which forms the backbone of each was continuous across
what is now Bournemouth Bay. Thus The Foreland (Handfast Point) corresponds
to The Needles, and the chalk ridge which runs through Corfe to White Nothe
separates, like its counterpart in the Isle of Wight, Tertiary rocks to the north and
Jurassic rocks to the south. The former make the low ground of Poole Harbour and
the Frome valley; the latter make the varied and beautiful coastline from Durlston
Head to a little west of Lulworth Cove.

Along the south coast the Jurassic rocks, clays, sandstones and limestones, are
nearly horizontal near St. Alban's Head, but are tilted northwards at high angles
farther west, in which direction they rapidly thin out. As a consequence partly of
erosion, partly of change of level, the outer fringe of Jurassic rocks is locally cut
through to form beautiful features such as Worbarrow Bay, Lulworth Cove and

F

Stair Hole. To the east of Worbarrow the Kimmeridge Clay forms ledges; to the west of Lulworth the chalk ridge comes to the sea, but locally is fronted by reefs of Jurassic rocks, including the Blind Cow off Swyre Head and the Bull off Scratchy Bottom. The chalk forms fine cliffs, reaching 500 feet at White Nothe. Locally there are landslips. The structure of all this stretch of coast is remarkably interesting and many details are dependent for their form on the interrelationship of structure and erosion.

The cliff top scenery is noteworthy throughout. Parts of the coast east of Lulworth are difficult of access, since a large area is owned by the Ministry of Defence. Much of this stretch of coast has strong associations with Thomas Hardy. There are also many features of archaeological interest, such as Flowers Barrow hill fort. The ecology of the area is also interesting. Studland Heath is a National Nature Reserve. The Dorset coastal footpath runs through the area.

11. CHESIL BEACH

SY 683735—SY 490887. Length 15·2 miles.

This beach is unique in these islands, and is continuous from Bridport to Portland. The shingle of which it is composed, more than 90% flint, is remarkably well graded from small pebbles less than a walnut in size at its north-western end to pebbles two to three inches, or even more, in diameter at the other end. The width and length of the beach also increase to the south-east. It is fed from both ends, but despite the great amount of scientific investigation to which it has been subjected, there is still no completely convincing explanation of the grading of the pebbles. From Bridport to Abbotsbury it is in contact with the land; from Abbotsbury to Portland a lagoon, called The Fleet, is enclosed.

Chesil Beach is of outstanding physiographic interest and is followed by a loop of the Dorset coastal footpath. The Swannery at Abbotsbury, which is well known, lies nearby.

12. LYME BAY

SY 420916—SY 134874. Length 20·0 miles, of which 2·75 miles are substantially developed.

The part of the bay included in this area extends from Sidmouth to Seatown, near Chideock, and spans a wide range of Mesozoic rocks, from the red Triassic sandstones in the west to the Jurassic clays and sands in the east. In the midparts there are important outliers of Cretaceous rocks (Greensand, Gault Clay, and Chalk). The deep cut valleys in the cliffs near Sidmouth, the chalk cliffs and landslips at Beer Head, the famous landslips at Dowlands near the Devon-Dorset boundary, and the well-known line of cliffs between Black Ven and Golden Cap, make this part of the coast of unusual significance not only to the geologist and physiographer, but to all who are interested in scenery.

The landslips at and near Dowlands are included in the Axmouth-Lyme Regis Undercliffs National Nature Reserve, and in the great gulley formed by the slip an ash wood has grown which is of much interest to botanists. There are several resorts along this stretch of coast, and the coastal footpath affords reasonable access to the cliffs. There are associations with Jane Austen at Lyme Regis, where the Harbour (The Cobb) is also of interest.

13. SCABBACOMBE HEAD

SX 925532—SX 893503. Length 4·4 miles.

The coast between Brixham and the mouth of the Dart is remarkably unspoiled.
It is a coast of steep cliffs and coves often difficult to reach from the land. The rocks
are Lower Devonian in age, mainly grits and shales strengthened by igneous intrusions. The different types of rock, the structure of the area, marine erosion and the
effects of changes of level of land and sea have produced a coastline of great beauty
and intricacy. The mouth of the Dart is an excellent example of a drowned valley.
Part of the area is a Site of Special Scientific Interest. The Devon South Coast footpath
follows the cliffs.

14. SOUTH DEVON

SX 818383—SX 675397. Length 16·0 miles.

Between Bolt Tail and Start Point, including the drowned valley of Salcombe
Harbour, is a beautiful stretch of coast, much of which is owned by the National
Trust. The rocks of which it is formed are highly metamorphosed schists of uncertain
age. They are strongly folded. The cliffs from Bolt Tail to Bolt Head are mainly
in mica-schist, a rock which also forms the craggy outline of Start Point. The remaining parts of the cliffs are in the green schists.

A prominent feature of the coast near Prawle Point and Lannacombe is the raised
beach which usually appears as a rock platform. At the back of the platform there is
often a low cliff in a deposit called Head. The pinnacles at Matchcombe were cut by
erosion when the sea stood at a higher level than now. The Head once covered the
whole platform.

The Devon South Coast footpath follows the cliffs and offers superb views. Bolt
Head also forms a fine viewpoint and is the site of a promontory fort.

15. RAME HEAD

SX 442487—SX 416503. Length 3·7 miles.

Rame Head is part of the peninsula on the western side of Plymouth Sound,
followed by the Cornwall South Coast long distance footpath. The headland rises to
about 300 feet, and is made of grey slates interstratified with a quartz-veined grit
which may explain its prominence. It is joined to the rest of the peninsula by a low,
narrow neck of land. There are remains of a Chapel and an ancient cliff castle. The
headland forms a good viewpoint.

16. POLPERRO TO GRIBBIN HEAD

SX 224515—SX 092520. Length 12·1 miles, of which 0·25 miles are substantially
developed.

This area lies in the Lower Devonian rocks which make fine cliffs between Polkerris
and Talland Bay, followed by the Cornwall South Coast footpath. Sandy beaches are
small and scarce, but the cliff scenery is of a high order. The River Fowey is an
example of a drowned valley. The platform of the raised beach is locally prominent.
Much of this part of the coast is owned by the National Trust, and Polperro Cliff
is a Site of Special Scientific Interest, on account of its botanical and ornithological
importance. The village of Polperro is picturesque and well known.

17. MEVAGISSEY TO ZONE POINT

SX 018437—SW 851325. Length 23·6 miles, of which 0·5 miles are substantially developed.

This stretch of coast is in Devonian rocks and is varied in nature. For the most part it is a cliffed coast, and there are two prominent headlands—Nare Head and Dodman Point. The former is a mass of igneous rock, the latter is in metamorphic rocks, largely slates and phyllites. The minor points are nearly all in resistant rocks, often of igneous origin. Kiberick Cove lies between the igneous masses of Nare Head and the Blouth, and the inlet of the Straythe lies between the Blouth and Manare Point. Jacka Point, Hartriza Point, and Caragloose Point are all igneous. Black Head, north of Mevagissey, is similar. The offshore rocks, e.g., Gull Rock, are nearly all igneous masses.

In the coves, many of which are cut along faults or are in closely folded rocks, there are small beaches, and larger ones at Gorran Haven and Pendower. The National Trust have several properties along this stretch of coast, which may be explored from the Cornwall South Coast footpath which follows its length. There are good views to be had from the prominent headlands.

18. LIZARD

SW 722145—SW 634250. Length 15·7 miles, of which 1·0 mile is substantially developed.

The surface of the Lizard area is a flat and somewhat featureless plateau about 250 feet high. The cliffs, on the other hand, are of much interest and beauty. The predominant rock is serpentine, three types of which can be distinguished. On the west coast Mullion Cove is in the serpentine, but a mass of schist forms Predannack Head. Serpentine cliffs soon reappear, but a little south of Kynance Cove give way to the schists which form Lizard Point and continue northwards to Landewednack.

Many of the coves are cut along small faults or where dykes of igneous rock reach the sea. The western cliffs are steeper than those to the east, a factor partly explained by the difference of exposure to the open Atlantic and the more sheltered waters to the east. In general the cliff scenery on the east is less imposing. North of Polurrian Cove, Devonian rocks reach the coast, which is indented and picturesque. The cliffs soon drop down to the sand and shingle bar which dams back the Loe.

The National Trust own several parts of this stretch of coast which is traversed by the coastal footpath. Lizard Point, being the most southerly point on the mainland of Britain, is popular with visitors. There is a lighthouse and a lifeboat station.

19. WEST PENWITH

SW 469258—SW 513410. Length 33·5 miles, of which 0·5 miles are substantially developed.

This rather unfamiliar name covers the whole of the great granite promontory of west Cornwall and includes the coast from St. Ives to Mousehole. Not all of the coast is granitic: St. Ives Head and Gurnard's Head are greenstone formations. From St. Ives to Morvah the coast is crenulate; there are many coves, and the cliffs are not particularly steep. Inland there is a bleak plateau. Between Pendeen Watch and Cape Cornwall is the finest example in England and Wales of cliffs cut in metamorphic rocks. There are numerous clefts or *zawns* in these cliffs.

From Cape Cornwall to Mousehole the cliffs are granitic. Their forms depend greatly upon the nature and inclination of the joints. From Cape Cornwall to Land's End there are high sloping cliffs, and there is a good beach at Sennen. The castellated form of cliffs is noticeable at Land's End, but becomes more pronounced in the magnificent coast between that point and Penberth Cove. Farther east the cliffs are less vertical.

The 'heritage' associations of Land's End, one of the major extremities of Britain, are a great attraction for visitors. Also of interest are the derelict engine houses associated with the former mining industry, especially near St. Just, and the Minack cliff-face theatre at Porthcurno. Archaeological remains of many kinds are abundant. The whole coast is of interest to geologists and physiographers, and certain parts of it to botanists. The coastal footpath follows the cliff line throughout. Although the coast road lies within a mile of the sea in most parts, there are many comparatively remote stretches, for example between Morvah and St. Ives.

20. ISLES OF SCILLY

Whole island group. Length 40·0 miles.

There are about 140 islands and skerries; 5 are inhabited. Four of the islands exceed 100 feet. They are in origin similar to the Land's End peninsula and other granite areas in Cornwall. In the islands the granite is, however, partly submerged. Sand bars often link two islands together. The granite breaks down into coarse sand, and in places there are curiously shaped eminencies consisting of great granite boulders weathered out along joint planes. Terraces are common, and are mainly formed of Head. The sand is spread over the surrounding sea floor and beaches and is subjected to considerable wind action. On Tresco, blown sand impounds a lake of fresh water.

The islands, and more especially the Seven Stones, are associated with the Legend of Lyonesse. There is much to interest archaeologists, biologists and physiographers. The cultivation of flowers adds much to the beauty of the islands.

21. TINTAGEL TO WIDEMOUTH

SX 044862—SS 199028. Length 19·4 miles, of which 0·1 miles are substantially developed.

The southern part of this piece of coast, from Tintagel to Boscastle, is in Devonian rocks. At Tintagel the structure is complicated. Tintagel is almost an island and there is a major fault between it and the mainland. The cliff scenery is fine and varies with the several rock types which reach the coast. The flat top of Tintagel Island is part of a much more extensive flat (the Trevena level) which can be traced along the coast. The phyllite (slate) cliffs in Bossiney Haven are noteworthy. The Valency (Boscastle Harbour) is a drowned valley with precipitous and craggy sides.

Immediately to the north the Devonian gives place to the Carboniferous. The range of cliffs from Boscastle through Cambeak and Dizzard Point to Foxhole Point shows infinite detail; their slopes are often tumbled and grassy, and offshore there are many stacks and islets. The folding in the cliffs at Millook is spectacular. The cliffs end at Widemouth, and give place to a boulder-strewn beach.

There are extensive National Trust holdings along this stretch of coast, which is also followed by the Cornwall North Coast footpath. The associations with King Arthur at Tintagel add interest. There are a number of good viewpoints.

22. HARTLAND

SS 202116—SS 317251. Length 19·6 miles.

The coast from Duckpool to Clovelly affords probably the grandest cliff scenery in England and Wales. The cliffs are high and broken by many small gorges made by streams draining to the west coast from a watershed three or four miles inland. The rocks involved are of Carboniferous (Culm) age, and are much folded so that in places the strata are sometimes vertical, in others they form arches (anticlines) or troughs (synclines). A marked feature is the number of coastal waterfalls. Sometimes these fall almost vertically, e.g. Litter Water, from a valley in the cliffs; others are sloping, and one or two are more complex. Milford Water, a fairly big stream, makes five distinct falls, controlled largely by the structure of the rocks. At Hartland Quay and Screda Point it is easy to see that marine erosion has cut into the valley of the Wargery Water which formerly drained to the sea just north of Hartland Quay. It now falls into the sea just north of St. Catherine's Tor (See also the Valley of the Rocks at Lynton, under Exmoor (23.), below).

The coast is fronted by a rocky platform which was mainly formed by the sea when it stood at a slightly higher level relative to the land. The cliff face shows how the sea can cut out ridges, pillars and small caves. If one looks down on to the platform from the cliff top the structure of the rocks is seen as if laid out as in a geological model. Higher and Lower Sharpnose Points afford magnificent views of the cliffs. The coast east of Hartland Point is rather less rugged since it is a little more sheltered. Near Clovelly it is thickly wooded. The street of Clovelly, now popular with holidaymakers, was once a watercourse.

The path along the cliffs (part of the coastal footpath), affords some of the finest coastal views in Britain. Lundy Island is conspicuous on a clear day. Features of interest include the headland of Hartland Point, with its lighthouse, and the vicarage at Morwenstow, with its associations with Vicar Hawker.

23. EXMOOR

SS 542485—SS 864481. Length 23·1 miles, of which 1·0 mile is substantially developed.

The coast from Rillage Point, near Ilfracombe, to Porlock Weir forms nearly all of the seaward edge of Exmoor and consequently of the Exmoor National Park. It is a high and beautiful coast but to appreciate much of its detail it should be seen from the sea. This is because the ground makes a convex slope down to the sea and only the lower parts are eroded by the waves. Hence, it is often difficult to see the true marine cliffs from above. There are several small streams which cut deep valleys. Heddon's Mouth is the best example; the joint mouth of the East and West Lyn rivers at Lynmouth forms a small harbour. It is difficult to realise that in the great rain storm of 1952 the combined rate of flow of the two streams for a few hours was almost as great as the record figure for the Thames. The Combe Martin valley is less attractive. The coast nearer Porlock is well wooded so that access to, and views of, the cliffs are often difficult. The Valley of the Rocks at Lynton is a dry valley; the effect of weathering is plain on the seaward wall which is formed of rather soft sandstone.

This coast is relatively sheltered and stands in strong contrast to that south-west of Hartland Point. The seaward slopes of Exmoor are land slopes and the true marine cliffs form a relatively small part of them. They are hog's back cliffs in contrast to the flat topped cliffs near Hartland and in many other parts of the coast. Magnificent views can be seen from Foreland Point, Trentishoe, and from high points on either side of Combe Martin Bay. All of these are accessible from the North Devon and Somerset coastal footpath.

The coast is of interest to the naturalist and botanist, especially at Heddon's Mouth, which is protected by the National Trust. There are also some fine woodlands, many ancient remains, including a Roman Signal Station, and a lighthouse at Foreland Point.

24. SOUTH GLAMORGAN

SS 973669—SS 847770. Length 11·9 miles.

There are two distinct and unlike parts in this stretch of coast. On the north and west side of the Ogmore River, at Merthyr Mawr Warren, there is an area of dunes. Blown sand hereabouts has been present since the Beaker Period (2000–1500 BC); it increased in the Middle Bronze Age and later. Despite some fixation in the Norman period and earlier, there was renewed movement in the sixteenth century, and between 1514 and 1575 it encroached seriously on the river. There have also been later movements. There is a fairly close connection between sand movements and archaeology and history, a connection found in many dunes on the coast of South Wales.

South of the Ogmore there is a cliffed coast. The rocks forming them are Mesozoic in age, but at Southerndown the actual foreshore is in Carboniferous rocks, and the cliffs in the Mesozoics. The cliffs run on to Nash Point and Cwm Col-hugh. These Liassic limestone cliffs are cut by several small faults.

There are several features of archaeological interest, including cliff-top camps, ditches and some more recent castles at Dunraven, St. Donats, Ogmore and Candleston. There is a beach at Dunraven and a lighthouse at Nash Point.

25. GOWER

SS 593876—SS 520954. Length 34·2 miles, of which 0·25 miles are substantially developed.

Most of the Gower peninsula is a plateau of Carboniferous Limestone which is folded in complex fashion. The synclines in Millstone Grit shales are mainly responsible for Oxwich and Porteynon Bays. There is also much faulting, some parallel with the folding, but most at right angles to it. Nearly all the minor indentations on the coast between Porteynon and Worms Head are eroded along small faults. Faulting is also present in Caswell Bay, Brandy Cove, Pwlldu Bay and Three Cliffs Bay.

The cliffs on the south coast show five main features—the level plateau surface, the steep fall to the remnants of the raised beach, the raised beach platform, the fall to the beach, and the present platform, which is wide and not wholly of modern origin. In Rhossili Bay there is a marked platform at about 100 feet, banked against the Old Red Sandstone of Rhossili Down. The origin of the platform is uncertain; it is not a raised beach.

The north-western extremity of the peninsula is a long sand spit, Whitford Burrows. To the east there are extensive marshes, between Llanrhidian sands on the north, and the old cliff on the south. This part is famous for its bird life.

The National Nature Reserves at Oxwich, Whitford and the Gower Coast, the Sites of Special Scientific Interest, and the large holdings of the National Trust emphasise the importance of the Gower peninsula in the conservation of the coast of England and Wales. Burry Inlet is of international ornithological significance.

There are also many ancient monuments and earthworks and some interesting caves. The prominent headland of Worms Head and the superb sweep of Rhossili Bay are much visited.

26. SOUTH PEMBROKESHIRE

SS 125983—SM 871029 (including Caldey Island). Length 41·0 miles, of which 0·5 miles are substantially developed.

This and the two following areas are included in the Pembrokeshire Coast National Park. The coastal footpath connects all three areas.

In few parts of the coast is the relation between structure and coastal scenery so plain. Between Linney Head and Broad Haven the Carboniferous Limestone makes a nearly level plateau, 100–160 feet high. The cliffs are usually vertical and many of the interesting features, caves, stacks and arches, are produced by marine erosion, but erosion working on a limestone mass already honeycombed by water courses formed at an earlier time.

The Old Red Sandstone cliffs occur at either end of an anticline which reaches the sea in Freshwater West Bay, and also in the east between Barafundle Bay and Skrinkle Haven. Several well known inlets, including Swanlake and Manorbier Bays, are cut along faults. Between Freshwater East and Old Castle Head the strata are nearly vertical and give a striped appearance where beds of different colours reach the surface of the cliffs or beach platform. In the west the headland north of Freshwater West Bay is in the Old Red Sandstone. The Carboniferous Limestone reappears between Skrinkle Haven and Tenby, and Caldey Island, on which there is a monastery, is a severed part of the mainland. The island is wholly formed of limestone. Neighbouring St. Margaret's Island is a reserve of the West Wales Naturalists' Trust, important for its bird life.

The whole peninsula is an area in which geologists and physiographers find great scope for their work, and because the relation between rock type and scenery is so clear it is an admirable field laboratory. The many sandy beaches are popular, and features such as Elegug Stack, St. Govan's Chapel and the Green Bridge of Wales are of interest. The Castlemartin Artillery Range of the Ministry of Defence occupies much of the area.

27. MARLOES AND DALE

SM 807028—SM 852128 (including Skomer and Skokholm Islands). Length 24·6 miles.

The south coast of St. Brides Bay shows cliffs cut in Carboniferous and Old Red Sandstone strata, and particularly imposing is an intrusive mass which forms Borough Head and Ticklas Point; the Stack Rocks are also part of it. Between Mill Haven and Musselwick the deep red of the cliffs is striking, and the many minor inlets are nearly all associated with faults. Older rocks (Ordovician) occupy a small area at Musselwick Sands. The blackness of these rocks is in marked contrast to the deep colour of the Old Red.

The relations between rock hardness and minor faulting are shown in the little peninsula running out to Midland and Skomer Islands, both of which are almost wholly formed of volcanic rocks. Skomer Island is a National Nature Reserve famed for its bird life. Skokholm has a breeding colony of Grey Seals and there is a bird observatory. Gateholm Island is a great stack of Old Red Sandstone, not yet quite cut off from the mainland. Marloes Bay is remarkably fine. There are many stacks and considerable variations in rock type. Dale peninsula is almost an island; it is separated from the mainland by a deep valley which may be associated with a great fault which crosses Milford Haven and reappears near Tenby. Milford Haven is a magnificent example of a *ria*, a type of sea inlet which runs more or less parallel with the trend of the folding of the rocks.

Other features of interest include a number of ancient promontory forts and settlements, as well as a lighthouse at St. Ann's Head and a number of other good viewpoints. The Pembrokeshire Countryside Unit is situated nearby, at Broad Haven, just beyond the northern boundary.

28. NORTH WEST PEMBROKESHIRE

SM 826229—SM 955394 (including Ramsey Island). Length 48·9 miles, of which 0·25 miles are substantially developed.

This long stretch of coast affords probably the most varied and interesting cliff scenery in England and Wales. It is less grand than that near Hartland, (22.) but shows a much wider range of rock types.

The lichen and plant covered cliffs, mainly cut in Cambrian rocks, are remarkably beautiful near St. David's and Solva. Solva Harbour and Porth-clais are good examples of drowned valleys, and the bays, St. Non's, Caerfai and Caerbwdi, are picturesque. The cliffs are steep and range in colour from purples to yellows and greys. There are many stacks and islets. The softer cliffs at Porth-y-rhaw indicate something of the rate of erosion: an old cliff camp is being destroyed by the sea. The west facing coast is perhaps less striking, but Ramsey Island (a detached part of the Pembrokeshire plateau) and Whitesand Bay give it great character. In Whitesand Bay and Porth-melgan glacial drift reaches to and below sea level.

From St. David's Head to Fishguard igneous rocks intruded into sedimentaries play a major role and make this stretch of coast of unique interest. Trwyncastell is formed of rhyolites, Abereiddy Bay and Traethllyfn are cut in dark shales, and the headland north of the Traeth is diabase. Between Trwynelen and Trwyn-llwyd, and also at Pwllwhiting, the coast is formed mainly of shales, but Castell-coch, Ynysdeullyn and Ynys y Castell are igneous masses. Even in minute detail there is an astonishing parallelism between the coastal crenulations and the variations of rock type. Strumble Head is another great mass of igneous rock.

Much of the interior of this part of Pembrokeshire is flat or pleasantly undulating; Carnllidi near St. David's Head is nearly 600 feet high, and dominates that part of the county. There are many ancient camps and promontory forts, and a rich ecclesiastical history based on St. David's Cathedral. Included in this stretch is the scene of the last sea invasion of the British Isles, Carregwastad Point, where the French landed in 1797. The National Trust hold many parts of this coast, which is also of interest for its bird life and its Atlantic Seal colonies.

29. ST. DOGMAELS

SN 110547—SN 164512 (including Cardigan Island). Length 8·9 miles.

This short stretch of coast includes Cemaes Head, the estuary of the Teifi and Cardigan Island. Part lies within the Pembrokeshire Coast National Park and is followed by the coastal footpath.

The folding in the cliffs near Cemaes Head is spectacular, and there is a fine range of cliffs as far as Pen-yr-afr. Just south of that headland is a valley, mainly dry, which runs from Pwllygranant to the Teifi estuary. This is an overflow valley, formed at a certain stage of the ice age by melt waters from the ice sheets. The Teifi estuary is a drowned valley, almost closed by a sandbar. It has a rich bird life and is popular with holidaymakers. Cardigan Island is a West Wales Naturalists' Trust reserve.

30. LLEYN

SM 317280—SM 296412 (including Bardsey Island). Length 44·4 miles, of which 0·1 miles are substantially developed.

Between Porth Dinlläen and Aberdaron Bay there is a coast of pre-Cambrian rocks. There are cliffs, but for the most part they are low. Glacial drift fills, to well below sea level, old bays at Porth Nevin and Porth Dinlläen. Dinlläen headland is made of igneous rock. Farther to the south-west the cliffs are low, partly in igneous and partly in sedimentary rocks, and locally, as at Porthoer, interrupted by glacial drifts. Along much of this part there is a belt of lower ground adjacent to the shore. The cliffs rise much higher and are beautiful near Braich Anelog and Braich y Pwll.

Aberdaron Bay lies in Ordovician rocks, but the high cliffs at its head are cut in glacial drift and show characteristic erosion features. The islands, Ynys Gwylan-fawr and Ynys Gwylan-bâch, are parts of an igneous sill which reappears on the mainland. Porth Neigwl, a dangerous bay for bathing, is filled with glacial deposits. The little Soch river almost certainly at one time ran into the bay; now it turns back on itself and cuts a deep gorge to run out to the sea at Abersoch. The St. Tudwal's peninsula is made of sandstones and flags of Ordovician age, and its two headlands enclose Porth Ceiriad.

The whole of this district is remarkably interesting; it is an area of pleasant scenery and one in which it is difficult to separate at all rigidly inland from coastal scenery. Bardsey Island and the Gwylan Islands are of ornithological importance. Many of the headlands, such as Mynydd Mawr, offer good views. There are several small resorts in the area, and a number of fine beaches.

31. HOLYHEAD MOUNTAIN

SH 243796—SH 224837. Length 8·0 miles, of which 0·5 miles are substantially developed.

This forms the highest part of this coast of Anglesey, rising to 720 feet at Holyhead Mountain, ¼ mile inland. Fine cliffs, often nearly vertical and showing much folding, are cut in quartzites of pre-Cambrian age. To the south as far as Porth-y-post the coast is lower, but beautiful. There are numerous crenulations and coves. The highest cliffs are at North Stack and South Stack; there is a prominent lighthouse at the latter point. This lighthouse, approached by a steep and tortuous path, and the shattered face of Holyhead Mountain, are among the features of interest.

32. NORTH ANGLESEY

SH 300892—SH 440936. Length 19·3 miles, of which 1·25 miles are substantially developed.

This is a picturesque coast. There are some fine bays, and the setting is pleasant. There is, however, no spectacular scenery. In Cemlyn Bay there is a good example of a mid-bay bar. Cemaes Bay, Hell's Mouth, Porth Wen, and Bull Bay are all good features.

Carmel Head stands in a relatively remote part and is rather higher than the coast to south and east. The western part of it is formed of gneiss, and a narrow band of Ordovician rocks reaches the sea near the Head. Dinas Gynfor, one of the prominent headlands, is the most northerly point of Wales and contains an Iron Age fort. There are many other historic sites in the vicinity, and a number of good viewpoints.

33. GREAT ORME

SH 768823—SH 782832. Length 4·4 miles, of which 0·5 miles are substantially developed.

The Great Orme is a mass of Carboniferous limestone nearly 700 feet at its highest point. It is now joined to the mainland by an isthmus of blown sand. There are some fine cliffs, especially on its north side, and a number of caves. The entire headland is traversed by a scenic road, the Marine Drive, providing access to the lighthouse and many other viewpoints. Being so close to Llandudno, the headland is popular with visitors.

34. ST. BEES HEAD

NX 959118—NX 961162. Length 4·2 miles.

Between Whitehaven and St. Bees the coast not only projects seaward, but rises in height to form the headland known as St. Bees Head. There are, in fact, two heads, North Head and South Head, separated by a small valley which drains to Fleswick Bay. The rock of which they are formed is sandstone of Triassic age. The cliffs are good, often falling sheer for 200–300 feet. On the beach in Fleswick Bay gem stones of some beauty are found. Near South Head differential erosion has produced fissures; Pattering Holes is the best known example. St. Bees Head is the most conspicuous feature on the whole coast between the Solway and North Wales. There is a prominent lighthouse on North Head.

APPENDIX 2.
Management Proposals for a Heritage Coast

This Appendix, which should be read in conjunction with the eight accompanying maps (following page 99), illustrates the application of the principles contained in this report to a typical stretch of coastline of high quality scenery. The maps represent a potential area of Heritage Coast,* and show how it might be planned and managed.

Map 1 shows the general features of the area. Maps 2, 3, 4 and 5 show the resources of the area, their recreational potential and ecological characteristics, the pattern of access, restrictive factors, and special opportunities available for management. Map 6 summarises all the factors mentioned and shows how these accord with the principles set out in the report, and particularly those contained in Chapter 6 concerning the management plan for a Heritage Coast. Maps 7 and 8 together indicate the proposals for the planning and management of the area.

The eight maps and these notes accompanying them illustrate the stages envisaged in making a plan and management proposals for any Heritage Coast, although local considerations will determine the final form of the plan. Clearly these principles can also be applied to other coastal areas which are not Heritage Coasts. The present analysis is perhaps little more than a rudimentary study of the resources of the area shown, and for the reasons mentioned in Chapter 6 a more sophisticated approach which takes greater account of the special characteristics and carrying capacities of these resources would be desirable.† However, as mentioned in paragraph 6.9, little progress has been made in establishing suitable methodology for this purpose so that the techniques demonstrated here may be of interest.

Three further points should be noted concerning the maps. Firstly, for simplicity the scale of 1 inch to 1 mile has been chosen; the 1/25,000 scale could however be used, and careful attention would, of course, have to be given to the Development Plans Manual.‡ Secondly, for clarity the maps are shown in colour, but monochrome notations could be used to reduce costs. Finally, the 4-figure and 6-figure numbers in the following notes which accompany the maps are National Grid map references and refer to specific features shown on the maps.

---

* i.e. an area selected according to the criteria set out in Chapter 4.

† See for example the report *East Hampshire Area of Outstanding Natural Beauty: a study in Country-side Conservation*, which illustrates the principles underlying resource zoning.

‡ *Development Plans: a manual on form and content*. M.H.L.G. 1969.

MAP 1. THE STUDY AREA

This map shows the main physical features, roads and settlements* along the stretch of coastline under consideration and in the adjoining area inland. The topographical detail shown on this map (with the exception of the contours) is reproduced in grey on Maps 2 to 8. In most cases the additional material shown on these subsequent maps is confined to within 2 miles of High Water Mark: this is the Study Area.

The greater part of the Study Area is substantially undeveloped. There are two small towns at its extremities, but neither would qualify as a resort. The coastline between these settlements is of great beauty and comparable to the 34 areas selected as described in Chapter 4. Several types of coastal scenery are represented.

MAP 2. RECREATIONAL RESOURCES

This map shows the recreational facilities and attractions of the Study Area and the quality and characteristics of the beaches. Although no measure is given of the degree of recreational use or popularity,† the range of existing services and facilities‡ is shown. The map shows the suitability of the Study Area for recreational activities based on natural features, and also indicates, for those parts where there is little recreational use at present, the recreational potential. The ecological significance and scientific importance of the resources is shown on Map 3.

The term 'open country' used on the map implies uncultivated land (including rough grazing) which is potentially suitable for public access.§ Such land may not necessarily accord exactly with the statutory definition of open country given in s.59 of the National Parks and Access to the Countryside Act 1949 (as amended by s.16 of the Countryside Act 1968). The extent to which public access is allowed is indicated on Map 5.

Each beach has been assessed in relation both to its size and its suitability for recreation. The extent of beach exposed at mean tide level is represented on the map by a circle of proportional size. The recreational suitability of each beach has been assessed in accordance with the six factors listed below, and points have been awarded for each factor.‖ The total number of points for the six factors is expressed diagrammatically on the map by the green segment within each of the beach circles.

Factor 1: Quality of beach material¶

Proportion of sand greater than 80%	15 points
Proportion of sand between 61% and 80%	12 points
Proportion of sand between 41% and 60%	9 points
Proportion of sand between 21% and 40%	6 points
Proportion of sand 20% or less	3 points

---

* It should be noted that the settlements include only groups of buildings, and that isolated farms and houses elsewhere are not shown. Permanent caravan sites (but not camping sites or seasonal caravan sites) are included within some of the areas shown as settlements: these are distinguished on Map 5.

† This is shown on Map 4 in terms of the location of the principal car parking areas.

‡ The facilities shown correspond to those listed in paragraph 9.6.

§ See paragraph 8.27.

‖ Other factors may of course apply locally in other areas and should be taken into account.

¶ Expressed as a percentage of the total beach surface exposed at mean tide level.

Factor 2:   Degree of safety*

Safe for most types of water activity	9 points
Reasonably safe for average swimmers	6 points
Safe for strong swimmers only	3 points
Dangerous for swimming or other types of water activity	0 points

Factor 3:   Shelter

Reasonably sheltered on three sides	4 points
Sheltered from the prevailing wind only	2 points
Not sheltered	0 points

Factor 4:   Aspect

South-facing beaches	4 points
East or west-facing beaches	2 points
North-facing beaches	0 points

Factor 5:   Tidiness†

Less than 10% of beach covered by tide-deposited refuse‡	2 points
Between 10% and 24% of beach covered by tide-deposited refuse	1 point
25% or more of beach covered by tide-deposited refuse	0 points

Factor 6:   Additional qualities or special features

Addition of 1 point each for the presence of any of the
following features (maximum 2 points):
rock pools
abundance of shells and other marine life   **2 points (maximum)**
beaches specially suitable for surfing
caves
waterfalls

Deduction of 1 point for the presence of any friable cliffs
susceptible to rock falls, or unexploded missiles or other   **Minus 1 point**
local dangers

Maximum possible (Factors 1 to 6):   36 points per beach

Man-made features of recreational significance are also indicated on the map. These include pre-Roman earthworks (214683 and 274667), a Norman Church (211694), timbered houses (253666) and a row of old thatched cottages (349747). The ancient island fortress (350665) is also of considerable interest and in conjunction with the adjacent headland may be regarded as a feature more of regional, than local, importance. In addition, there are several interesting caves (around 213650), a logan or rocking stone (345709) and a prominent lighthouse (245632). Linear features include the track of a disused railway (from 334678 to 349757) and several prominent ridges and valleys which contain features of interest and could form the basis of interpretative trails and walking routes.

It will be seen that the range of existing visitor facilities is somewhat limited. There is only one information centre and few refreshment places or public lavatories. No properly laid out picnic sites, nature trails or scenic routes are shown. Sailing is

---

* Aspects considered include the depth of water, the offshore gradient of the seabed, and the presence of currents, submerged rocks, etc.

† Expressed as a percentage of the total beach surface exposed at mean tide level.

‡ Including litter originally deposited by visitors, and other tide-borne material such as seaweed, driftwood, oil, sewage, etc.

limited by the shortage of mooring facilities. The implications of these deficiencies are considered in Maps 6, 7 and 8.

Map 3. Ecological Characteristics

This map, used with the accompanying Table of Ecological Zones (page 88), shows the scientific significance of the Study Area and the interest and importance of its geological and physiographical features and its plant and animal life. The landscape types shown on Map 2 form the basis of the ecological zones, which are distinguished on the present map and are detailed in the accompanying Table. Limitations of scale have led to some generalisation and the exclusion of the smaller areas of semi-natural vegetation, including hedgerows and ditches. Nature Conservancy designations are indicated on the map. There are no local nature reserves in the Study Area.

The blue lines offshore represent the degree of scientific importance of each stretch of coast. Aspects which have been taken into account in arriving at this evaluation include the national or regional significance of rare species and the need to conserve specified habitats for purposes of education and research. This evaluation is shown in relation to the vulnerability of the coast under study, indicated by the red lines offshore. The latter evaluation indicates the degree to which the coast is vulnerable to ecological damage from a variety of causes—erosion, trampling, collecting of specimens, damage by fire, or other forms of disturbance. Both the blue and red offshore lines are based on information in the Table of Ecological Zones, which in turn is based on material supplied by the Nature Conservancy.

As will be seen, most of the Study Area is arable land or improved pasture; there are also several small areas of woodland and an almost continuous strip of semi-natural grassland along the cliffs and shore. Although cliffs predominate there are areas of salt marsh and sand dune and a number of bays, inlets and estuaries.

Map 4. Communications

This map shows the existing pattern of access by road, footpath and bridleway, and the main parking areas. It indicates the popular areas and the less accessible parts of the coast. Access points from the sea are also shown.

The first three categories of roads and motorable tracks shown are those where public rights of way exist for vehicles. Those shown as motorable tracks include unsurfaced lanes and smaller tracks on which it is possible, in reasonable weather conditions, to drive a private car. Some of these are shown on the definitive maps as 'roads used as public paths' and all are vehicular rights of way. In addition, *de facto* vehicular access may be said to exist on those roads and motorable tracks which are indicated as private but which are not shown to have gates or other physical barriers regulating entry. Most of these lead to farms and isolated buildings. On many of these roads and tracks new measures to restrict traffic, other than that of residents and service vehicles, will be necessary.

The parking facilities shown on the map indicate the number of pubilc car parking spaces throughout the Study Area, but do not include those reserved exclusively for specific facilities such as camping and caravan sites. The figures for the settlements indicate the total capacity of public car parks and authorised roadside parking areas. Outside settlements the spaces are either properly designed lay-bys, indicated as such by the highway authority, or other acceptable parking places where there are no amenity objections and where there is no obstruction or danger to traffic. The information shown is derived from low-level aerial surveys carried out on three average summer weekends, checked subsequently on the ground.

TABLE OF ECOLOGICAL ZONES

(To be read in conjunction with Map 3 and the notes on page 87)

Zone Numbers	Description	Particular features of Scientific or educational interest	Vulnerability
1–2	Well-drained, unsown calcareous grass-land above cliffs, mainly between 100 and 200 feet above sea level. Heavily grazed by sheep. Well used for walking, horse riding and car parking.	Characteristic herb species with occasional local rarities.	As concentration of recreational pressures is already causing the formation of gullies, paths should be diverted to allow the scars to heal, and alternative management techniques used to canalise visitors. Any significant decrease in grazing pressure would allow calcareous herbs to be swamped by rank grasses and scrub. Any substantial increase (e.g. by introduction of a large rabbit population), coupled with increased public use, would be likely to damage the turf. Agricultural reclamation even of part could reduce the area to a non-viable size and lead to the destruction of the habitat.
3–5	As for zones 1–2 but mainly above 200 feet and rising to 450 feet above sea level. Includes patches of acid terrace deposits and associated species. Well grazed but little recreational use.	S.S.S.I. designation reflects diversity of wild life, especially seabirds.	As for zones 1–2 but recreational pressures are as yet less significant.
6	As for zones 3–5 but lying inland, rising mainly to about 450 feet in places, with substantial areas of terrace deposits. Well grazed and a focal point for recreational activities on account of good access.	S.S.S.I. designation justified by diversity of herb species.	As for zones 3–5 but recreational pressures, and particularly indiscriminate car parking, are causing destruction of turf and reducing the size of the habitat. If this remains unchecked, the area will rapidly cease to be viable. Car parking should be confined to the edges of the zone.
7	As for zones 3–5 but little used agriculturally with consequent growth of gorse. Little recreational use.	S.S.S.I. designation on coastal sections on account of sea bird colonies and extent and diversity of plant species.	The gorse and other rough vegetation is rapidly expanding in area and control is needed if the turf is to be maintained. Grazing would be less disturbing to bird life than cutting or burning.
8–9	Acid water meadows and ley pasture with marshy patches. Unsuitable for extensive improvement but capable of grazing cattle during the summer.	Varied wildfowl. Useful teaching habitat, with wetland successions.	Natural succession and adjacent drainage works are drying out the marsh areas to the detriment of wildfowling and teaching interests.
10	Unsown, dry acid grassland behind sand dunes. Not used agriculturally.	No particular interest, but adjoins important nesting ground for terns (zone 22).	Vulnerability insignificant.

11–12	Mature, even-aged beech plantations with some ash edges. Not easily accessible except on foot.	Characteristic wild life species.	Recreational use no problem at present, but situation requires careful observation. Fire risk.
13–17	Mixed deciduous woodlands with occasional conifers and patches of hazel coppice. Much used by picnickers.	Important local wild life reservoirs within agricultural areas. Several used for school field studies.	Some danger of over-collection of certain species. Coppicing no longer takes place, with consequent loss of variety of habitat for both wild life and teaching purposes. A few places heavily used by picnickers may need enclosure to encourage woodland regeneration. Fire risk.
18–21	Conifer plantations with deciduous edging. Managed by Forestry Commission. Recreational use excluded.	No particular interest.	Fire risk. Zones 19 and 20 are mature and will shortly be felled.
22–23	Coastal sand dunes. Foredunes largely bare. Main dunes carry marram but with several blow-outs on paths to the beach.	S.S.S.I. designation. Clear physiographical and ecological succession for teaching purposes: sand movements, pH changes by leaching and vegetational adjustments. Important nesting sites for terns.	Some demand for sand removal by construction industry. Recreational use by vehicles, horses and walkers causing blow-outs where dunes are still accessible. These pressures should be controlled and diverted to firmer surfaces and away from nesting areas at critical periods.
24	Actively extending saltmarsh, partly protected by sand dunes.	Good teaching situation. Used for longer term university research on accretion rates and Spartina genetics. Wide range of wildfowl, including several national rarities which led to N.N.R. designation.	School parties with access permits occasionally disturb experimental equipment. At present no threat of reclamation for grazing. Wildfowl poachers in evidence but serious encroachments are rare. Motor boats, causing both erosion and disturbance to wildfowl, are a more obvious and serious threat.
25	Improved calcareous pasture in small hedged and walled fields. A few arable rotations. Mainly sheep and cattle.	Floristically rich but no rarities. Minor wildlife reservoirs provided by hedges and small scattered woodland clumps.	Many hedges and walls are not being maintained. Walls damaged by vandals in several coastal locations.
26	Chalk downland, mostly in large arable fields but with some short term ley pastures. Few hedges.	Interesting to geographers because of historical and recent land use changes.	Widespread use of chemicals restricts ecological significance.
27	Mixed arable and ley pastures with hedges and small groups of trees around large fields.	No particular interest other than minor wild life reservoirs in hedges and groups of trees.	Mechanical cutting of hedges prevents regeneration of hedgerow trees. Use of chemicals is an increasing threat to wild life.
28	As for zone 27, but with smaller fields and more ley pasture.	Habitat diversity produces local wildlife reservoirs.	No particular vulnerability unless agricultural practices intensify.

Zone Numbers	Description	Particular features of Scientific or educational interest	Vulnerability
29–40	Cliffs, wave-cut platforms, and rocks. Cliffs vary in height, up to about 350 feet above sea level (zone 33). Various limestone and chalk formations, some suitable for rock-climbing.	Some interesting physiographical features: wave cut platforms (zone 37), truncated stream valleys (zones 33 and 38), caves (zone 32), and landslips on chalk. Plant species related to adjoining zones. Many seabirds, especially in zones 32, 33 and 34. Geological interest limited by inaccessibility, but best observations made in zones 37, 38 and 39.	No particular vulnerability but occasional conflict between climbers or geologists and wild life, especially during nesting season. (Greater dangers are those to the public caused by steep nature of cliff faces.)
41–53	Simple bay head beaches of sand and shingle and stretches of coastal deposition.	Smaller bays of no particular interest. Zone 44 much used by field study parties interested in marine ecology. Zones 47 and 50 form part of succession with adjoining dunes, as does zone 51 with the salt marsh, which is slowly extending southwards over the mud flats.	Vulnerability insignificant, other than aspects mentioned in connection with zones 22, 23 and 24. Increased recreational use of the estuary (zone 44) should not involve dredging as this would seriously disturb marine life.

It should be noted that the figures for acceptable parking places indicate the capacity of the parking areas and not the extent to which they are used. A number of stretches of road are also shown where parking occurs but where it is undesirable. These include soft verges, passing places, spaces near driveways and entrances, and other locations which give rise to traffic or amenity objections. This information is based on the same survey but indicates *actual spaces in use* at these times; it may thus be assumed that in general other parking places are fully used and that an excess of demand over capacity exists on such occasions.

MAP 5. RESTRICTIONS AND OPPORTUNITIES

This map shows the background pattern of existing ownership and management against which the material shown on Maps 2, 3 and 4 must be considered. Where ownership is not specified, the land is privately owned. Special problems and restrictions are also identified and other classifications are noted.

It will be seen that most of the Study Area comprises privately owned agricultural land. However, a number of special management arrangements have been concluded between private landowners and public bodies. Woodland which is the subject of a Forestry Commission dedication scheme would normally be included also, but there are no examples in the Study Area.

The map illustrates the scope available for schemes of management. The restrictions and problems indicate where special measures may be needed to conserve the landscape or provide public access. The existence of publicly held land, or land subject to covenants or agreements, indicates where special opportunities may exist for establishing codes of practice or for co-ordinating management policies.* Other privately owned land, not specially indicated, may also of course present such an opportunity if the landowner's co-operation is sought. Land already developed or committed for development† is also shown; in the case of committed land this may suggest where revocation would be justified.

Four categories of restricted access are shown. Those which apply to enclosed farmland indicate areas where access would normally be confined to public footpaths, but none exist and, by reason of land use or ownership, the possibility of securing rights of way is virtually non-existent. Elsewhere over much of the Study Area access is also confined to public footpaths but it may be assumed that a reasonable possibility exists of diverting paths or creating new ones. Areas of open country indicated on the map as under threat of restriction include cases such as those mentioned in paragraph 8.29. Other areas of open country within the Study Area which are not shown on this map as subject to any form of restriction may be assumed to possess *de facto* rights of access.

Places where special problems arise are also distinguished. Eyesores and disfigurements shown include badly laid out caravan sites, areas of shack development, a car breaker's yard, unsightly advertisement hoardings and various military installations. Only the more important sites are recorded and there are many others where visual improvements are desirable. Measures for eliminating or reducing the impact of these disfigurements are included in the management schemes shown on Map 7 and described in the Table accompanying that map.

* See paragraphs 12.15 to 12.18.

† This includes not only land allocated for development in the development plan, but also (1) land which the local planning authority has, by resolution, determined should be so included, and (2) outstanding planning consents not yet implemented.

*G

MAP 6. APPRAISAL

This map summarises on one sheet the main elements of the survey material presented on Maps 2, 3, 4 and 5. It serves as a basis for analysis, and for the formulation of management schemes and proposals.

The popular parts of the Study Area are indicated by the main car parking locations and these correspond to the three categories of acceptable parking areas shown on Map 4. At the other extreme, the relatively remote and inaccessible parts of the area are shown, in terms of both the distance from public roads and the degree of 'apparent remoteness' they possess. The latter may be said to apply to those parts where, because of the ground cover or topography, an observer would not be aware of any buildings, traffic or other appreciable signs of human activity, and where a definite feeling of seclusion and remoteness could be obtained.

The classification of the coastline on this map is based on the evaluation of scientific importance and vulnerability shown on Map 3. The 'good beaches' indicated correspond to those (shown on Map 2) where the favourable characteristics equal or exceed two thirds of the possible total.*

The groupings of features of interest are based partly on the linear topographical features shown on Map 2, but also take account of antiquities and other features of countryside interest which occur in clusters and could be linked by interpretative trails.

Areas and sites requiring special management are also indicated. The restrictions, and sites for visual improvement, are based on those shown on Map 5. The areas of conflict include areas of ecological vulnerability† and certain places where other problems have arisen—for example, excessive car parking which requires to be controlled.

From an appraisal of these and other factors, and particularly from the general pattern of intensity, it should be possible to determine the basis of the management zones referred to in paragraph 6.16 which should form the framework of the overall plan for any Heritage Coast.‡ The most popular parts, or those suitable for large numbers, or those which contain facilities having a surplus capacity, or 'heritage' features of national or regional significance, suggest themselves as suitable for inclusion in an intensive zone. Those parts containing important ecological habitats vulnerable to damage, or relatively inaccessible stretches or secluded beaches, or which are suitable for low-intensity activities such as ornithology or cliff climbing, would fall most appropriately within a remote zone.§ Other factors summarised on this map which must be taken into account in drawing up boundaries include the pattern of existing settlements, eyesores and disfigurements, areas of restricted access and other problem areas, the extent of land in public or quasi-public ownership, and the distribution of features of interest.

The appraisal of the survey material shown on Maps 2, 3, 4 and 5 also indicates the scope for schemes of detailed management.‖ Thus the pattern of roads, footpaths and bridleways, in relation to the beaches, settlements and other principal focal points, indicates whether any changes in this pattern are necessary and where parking

---

* i.e. where the beaches have been awarded 24 points out of the possible maximum of 36, based on the six factors set out in the notes describing Map 2 in this Appendix.

† Defined on Map 3 and described in the Table of Ecological Zones.

‡ The degree of access and the scale of activity should of course be determined according to the acceptable optimum levels of use (see paragraphs 6.5 to 6.15).

§ The proposed management zones for the Study Area are shown on Map 8.

‖ Management schemes proposed for the Study Area are shown on Map 7 and are detailed in the Table of Management Schemes accompanying that map.

facilities should be provided. The pattern of roads and parking places, seen in relation to areas of pressure and conflict, also indicates where traffic management should be considered. The areas of ecological vulnerability shown on the maps indicate where special measures may be necessary to conserve certain habitats. Problem areas and restrictions shown may also suggest locations for schemes of positive action to secure access or landscape improvement, or to regulate the numbers, activities and movements of pedestrians. Groupings of features of interest likewise suggest scope for trails and interpretative measures. The lack of existing facilities and services, in relation to the focal points and management zones envisaged, indicates where new facilities should be provided. In relation to the other features of the area, this shortfall also indicates the scope for the introduction of facilities such as scenic routes, nature trails and picnic sites.

Map 7. Management Proposals

Maps 7 and 8 together indicate the proposals for that part of the Study Area which justifies designation as a Heritage Coast, and together they form the basis of an overall plan for the Heritage Coast. Map 7, used with the accompanying Table of Management Schemes, (page 94), shows the proposed schemes of detailed management, and also alterations in the pattern of access. Map 8 indicates the management zones, proposed new facilities, and other aspects of policy. On both maps the 'Boundary of Plan Area' represents the boundary of the area of Heritage Coast for designation purposes.

A number of changes are proposed in the pattern of communications. These are in accordance with the principles set out in Chapter 8, having particular regard to the implications of providing access in each of the management zones.* These proposals should be considered in relation to the existing network of roads, footpaths and bridleways shown on Map 4. The resultant pattern, taking these changes into account, is shown on Map 8.† Map 7 also shows proposed additional car parking areas, but these do not include parking facilities proposed in connection with schemes detailed in the accompanying Table of Management Schemes. The latter are however shown on Map 8.

An attempt has been made to link, by footpaths, principal focal points and features of interest, to provide a cliff or shore path with alternative return routes, to steer walkers away from fragile habitats and to make other minor adjustments such as new links and diversions. It is also proposed that a new footpath should follow part of the disused railway running north from 334678 and should continue beyond the Heritage Coast boundary.

The 21 areas proposed for schemes of positive management (which are based on the principles outlined in Chapters 8, 9 and 10, and on the need to conserve features of scientific interest) are bounded by a brown line on the map and are described in detail in the accompanying Table of Management Schemes (page 94). The management objectives for each scheme are indicated on the map by a code letter, and the methods by which it is proposed that these objectives should be achieved are shown

---

* For example: (i) The road south of 281678 is to be closed to motor vehicles to consolidate two stretches of relatively inaccessible coast and thereby to allow a larger motorless area to be managed as a remote zone; (ii) Three new short connecting lengths of road are proposed in order to create the scenic routes shown on Map 8; (iii) a vulnerable area in the vicinity of the National Nature Reserve is to be safeguarded by closing the short length of road north-east of 348721 and providing (for wildfowlers and casual visitors) a new car park further from the nature reserve, as well as a new footpath (see paragraph 8.3(h)).

† Roads which it is proposed should be restricted to certain classes or types of vehicle (see paragraph 8.3(i)) are however shown on Map 8 alone.

TABLE OF MANAGEMENT SCHEMES

(To be read in conjunction with Map 7 and the notes on page 93)

Scheme Number	Ownership	Principal Management objectives	Methods envisaged	Detailed proposals
1	Part District Council; part privately owned	(a) Control of pedestrians on headland. (b) Provison of adequate facilities for the enjoyment of this popular local beauty spot and viewpoint.	Part purchase; part consultation	1. Southern part of area to be zoned intensive (see Map 8). 2. Purchase of southern part of headland to consolidate public ownership and facilitate a comprehensive scheme of management. 3. New right of way for pedestrians to tip of headland. 4. Scheme for periodic re-alignment of footpaths within intensive zone to allow turf to recover. 5. Provision of new car park (10–20 spaces) to serve intensive zone. 6. Creation of picnic site on headland and nature trail along the footpaths.
2	Golf Club	Improvement of appearance of south-western portion.	Management agreement	1. Removal of massive and unsightly brick wall running from 218691 to 221689 which obstructs view up valley. 2. Replacement by low wall of local stone. 3. Limited planting. 4. Proposed new bridleway along line of new wall.
3	Common land Conservators	(a) Conservation of important local S.S.S.I. (b) Traffic control measures to prevent further erosion by vehicles. (c) Improvement to surroundings of earthwork at 234660. (d) Provision of new facilities.	Consultation	1. Provision of car park (30–40 spaces) to serve visitors to common land. 2. Restriction of vehicles elsewhere over the common by means of ditches, earth banks, etc. 3. Restriction of heavy vehicles on two roads through the area. 4. Provision of new bridleway and restrictions on horse-riding elsewhere over the common. 5. Creation of picnic site and nature trail close to new car park. (Trail to form part of longer route.) 6. Marking of paths to encourage their greater use, and provision of interpretative facilities at ancient earthworks at 234660. 7. Improved access to earthworks, including removal of scrub and rubbish dumped nearby.

No.	Owner	Objective	Action	Details
4	Private owner	Improvement of appearance of an area which is prominent from a number of surrounding points.	Discontinuance order	Discontinuance of use of land as caravan site, and return to grass. Owner to be given option of replacing caravans by a landscaped picnic site, which would otherwise fall within Scheme 5.
5	National Trust	(a) Conservation of local scientific interest along west coast. (b) Provision of facilities for visitors to headland and beach.	Consultation	1. Closing of private road to vehicular traffic would create need for new car park. This could be sited on Trust land at 217657. 2. If picnic site not provided as part of Scheme 4, this could be sited at 214667. Provision of anchorage facilities also at this point. 3. Creation of new stretch of coastal footpath, carefully planned to avoid vulnerable areas. Horse riding not to be permitted, but path could form part of nature trail.
6	Commons Management Committee	(a) Clearance of wartime eyesores on headland. (b) Provision of facilities in connection with proposed scenic route.	Consultation	1. Scheme for removal of wartime eyesores on headland at 2462, and restoration of land. 2. Provision of picnic site and car park beyond the lighthouse at 245632. 3. Provision of a series of small car parks or lay-bys on northern edge of common adjoining proposed scenic route. Restriction of vehicles elsewhere over the common. 4. Improved access for horse riders, with direct route from the stables at 251656. Riding to be excluded from the eastern part of the headland to minimise ecological damage.
7	Forestry Commission	(a) Improvement of access leading from valley to southern headland by inland route. (b) Provision of new facilities.	Consultation	1. Provision of car park (5–10 spaces) where plantation adjoins main road to north. 2. New footpath with trail to be made along western boundary.
8	Private owner	Improvement of appearance of land adjoining settlement.	Discontinuance order	1. Removal of unsightly shacks and old caravans by phased programme of demolition. 2. Consideration to be given to replacement of some accommodation as part of a properly designed scheme, with facilities for a sailing club incorporated.
9	Private owner	Improvement of appearance and elimination of land use incompatible with coastal conservation in a Heritage Coast.	Discontinuance order	Discontinuance of car breaking yard at 250652 and its replacement by a use appropriate to the area, or its return to agriculture.

Scheme Number	Ownership	Principal Management objectives	Methods envisaged	Detailed proposals
10	Two private owners	Improvement of landscape by removal of military eyesore.	Management agreement	1. Scheme for removal of eyesore at 261634 and restoration of site. 2. Co-operation of owner (who has already entered into an access agreement) to be sought to improve pattern of communications, including extension of bridleways. Riding to be controlled along those parts of the area where liable to ecological damage.
11	Private owner	Provision of access to open country.	Access agreement or order	1. Removal of barriers to access along cliff top by seeking agreement with owners. 2. Creation of new footpath, set back from cliff edge to avoid disturbance to wild life.
12	Private owner	Landscape improvement.	Discontinuance order	Removal of large unsightly advertisement hoarding on roadside.
13	Estate Company	Provision of new facilities.	Management agreement	1. New car park (5–10 spaces) just inside northern boundary as part of a series of car parks to serve the remote zone. 2. Opportunity to be sought to secure a right of way on foot along private road. (Owner has already entered into an access agreement.)
14	Several private owners	(a) Control of traffic and parking on edge of dunes and around settlement. (b) Control of pedestrians in dune area where there is erosion. (c) Conservation of dune environment and associated wild life. (d) Provision of new facilities.	Part by revocation; part by purchase; part by management agreement	1. This area should comprise a comprehensive management scheme implemented partly by local authority purchase and partly by management agreement. Part of the area to be zoned intensive to be acquired by the local authority to secure effective management of parked vehicles and access routes to the beach. 2. Revocation of outstanding planning permission for chalet development since further accommodation on this site would conflict with the management objectives for this area. 3. Site referred to in (2) to be used for additional car parking required, new public lavatories and first aid post. 4. All other parking, on road and in dunes, to be curtailed, except for second new car park to serve those wishing to visit quieter eastern part of area. Paths in this part to be carefully defined to avoid disturbance to wild life.

No.				
14 (Contd.)				5. This area could be used for experimental work on the carrying capacities of resources (referred to in Chapter 6) and for studies of visitor behaviour and wild life.
15	Ministry of Defence	(a) Improvement of landscape. (b) Provision of access.	Consultation	1. Agreement to be sought from Ministry of Defence on tidying up several barbed wire enclosures throughout this area, and particularly the removal or landscaping of unsightly target areas on the beach. These are prominent in the view from the headland at 344671. 2. Discussions with Ministry to proceed with a view to securing access to the beach when not in military use.
16	National Trust	(a) Provision of facilities. (b) Removal of eyesore. (c) Regulation of visitors to avoid further erosion on tip of headland, which is a heritage feature of regional significance.	Consultation	1. Creation of new footpaths from car park to popular viewpoint with rotation system to allow turf to recover. 2. Additional car parking facilities to be provided with corresponding restrictions on roadside parking to the north. 3. Replacement of unsightly refreshment shack by a new low block incorporating public lavatories and information services. 4. Improvement to surroundings of headland, which are both over-used and untidy in places, partly by recruitment of additional wardens. 5. Many of the proposals to be carried out jointly by National Trust and the local planning authority. 6. The landward part of the area to be zoned intensive.
17	Common land Conservators	Provision of new facilities.	Consultation	1. Provision of new picnic site and parking facilities at 340696. 2. Provision of two additional car parks (of 5–10 spaces) on edge of common for those using the scenic route. Parking elsewhere to be prohibited.
18	Private owner	Provision of access to open country.	Access agreement or order	1. Removal of barriers to open country and beach and provision of continuous strip of open country along coast. 2. Securing of agreement with owner not to plough headland in the north of the area.

Scheme Number	Ownership	Principal Management objectives	Methods envisaged	Detailed proposals
19	Part Nature Conservancy; part privately owned	Conservation of wild life and protection of National Nature Reserve.	Consultation	1. Consultations with Nature Conservancy and private owners regarding implementation of measures needed to protect the environment within the Nature Reserve, particularly the control of boating. 2. Investigation of possibility of opening the Reserve to interested members of the public at certain times of year.
20	Two private owners	(a) Conservation of wetland habitat. (b) Provision of facilities.	Management agreement	1. Proposed new picnic site and car park at 352739. This point provides a fine view of the marshes. 2. Consultation with owner regarding that part of the area adjoining the wetland habitat of the Nature Reserve, which could be adversely affected by any drainage scheme.
21	Several private owners	(a) Control of traffic. (b) Provision of facilities.	Purchase	1. This beach area close to the town should be purchased to enable parking to be controlled, partly by closing minor roads and using these for parking and partly by limiting the number of cars on land between parking areas and the beach. 2. The extension of the popular beach area further westwards towards the Nature Reserve should be curtailed by prohibiting vehicles west of 381733 and restricting access beyond this point to those on foot. A system of paths leading to the beach should be established within that part of the area to be zoned intensive. 3. Public lavatories to be provided within the intensive zone.

98

by the use of an appropriate colour. The implications of these proposals, and particularly of the management agreements envisaged, are considered in paragraphs 12.13 to 12.21. The schemes referred to attempt to resolve only the major problems requiring urgent attention; reference to Map 5 will indicate many additional areas where some form of action may be justified.

MAP 8. POLICY

This map, which should be read in conjunction with Map 7, summarises the policy which is intended to guide the overall planning and management of the Heritage Coast. Supplemented by the necessary policy statements, this map, together with Map 7, would correspond to the district plan referred to in paragraph 12.24. The area within the Heritage Coast boundary* is divided into the principal management zones described in paragraph 6.16. The factors underlying the selection of these zones and the determination of their boundaries have been outlined earlier in this Appendix in the notes which describe Map 6.

Intensive zones are proposed in four areas to cater for large numbers of visitors. These include the fine beach in square 3067 and the popular beauty spot on the headland at 348670, which, as already mentioned, is also a regional feature of some importance. Two remote zones are proposed: the National Nature Reserve and the relatively remote stretch of coast north east of 270631. The settlement zones include all the main settlements within the boundary of the Heritage Coast. The remainder of the area falls within the transitional zone.

The communications pattern shown on this map takes account of the proposed changes indicated on Map 7. Scenic roads† are also shown. It is intended that these routes should be associated with specially-devised trails linking features of interest.‡ Other interpretative techniques described in Chapter 11 should also be incorporated. Locations for parking facilities are also indicated; these include both the existing facilities (as shown on Map 4) and new proposals (shown on Map 7), and also those referred to in the Table of Management Schemes, which are not separately shown elsewhere. Road and parking place capacities are designed to relate to the desired overall level of use throughout the Heritage Coast.

Existing and proposed facilities and services for visitors are also shown, and correspond to those listed in paragraph 9.6. Their locations have been determined in accordance with the principles set out in the Facility Provision Chart. § Many are referred to in the Table accompanying Map 7 in connection with specific management schemes. A number of trail routes are envisaged in connection with the footpath system. It is intended that these trails should be varied in both nature and interest, in the ways indicated in paragraph 11.8.

* Shown on the map as 'Boundary of Plan Area'.
† See paragraph 8.13.
‡ See paragraph 11.9.
§ See page 45.

Printed in England for Her Majesty's Stationery Office by
Hindson Reid Jordison, Newcastle upon Tyne   Dd. 501278   K8

MAP 5   restrictions and opportunities

PUBLIC AND QUASI-PUBLIC OWNERSHIPS

Local Authority   `|||LA|||`

Ministry of Defence   `=MoD=`

Nature Conservancy   `/NC/`

National Trust   `\\NT\\`

Forestry Commission   `/FC/`

Common Land   `|||C|||`

PRIVATE OWNERSHIPS

Large areas in private ownership under unified management

Golf courses   `G`

EXISTING MANAGEMENT AGREEMENTS AND COVENANTS

Access agreements

National Trust covenants

Nature reserve agreements

DEVELOPED AND COMMITTED LAND

Land covered by buildings and associated uses

Permanent caravan sites

Land committed for future development

Seasonal caravan sites and camping sites

AREAS WITH RESTRICTED PUBLIC ACCESS

Restricted access over open country

Restricted access over foreshore

Restricted access over enclosed farmland with little or no prospect of establishing public rights of way

Areas of open country under threat of restriction by ploughing, fencing, etc.

SPECIAL PROBLEMS

Eyesores and other unwelcome visual intrusions   ●

Uses of land incompatible with the landscape and character of the coast   ★

Other locations where landscape improvement would be desirable   ▼

OTHER CLASSIFICATIONS

Area of Outstanding Natural Beauty

Article 4 Directions   `/////`

Ancient Monuments

MAP 8

policy

Boundary of Plan Area

MANAGEMENT ZONES

Intensive Zone		Remote Zone	
Transitional Zone		Settlements	

COMMUNICATIONS

Main roads		Scenic roads	
Secondary roads		Public bridleways	
Minor roads and motorable tracks open to the public		Public footpaths	
Private roads and motorable tracks		Roads restricted to vehicles of a specific type	
Gates and other physical barriers regulating vehicular entry to private roads and motorable tracks	×		

PARKING FACILITIES (OUTSIDE SETTLEMENT ZONES)

Less than 5 car parking spaces

Over 5 but not more than 10 spaces

Over 10 but not more than 20 spaces

Over 20 spaces

VISITOR FACILITIES AND SERVICES

	Existing	Proposed		Existing	Proposed
Restaurants (serving main meals)			Other moorings and anchorages		
Cafés (serving snacks and drinks)			Stables		
Information Centres			Boat trips		
Information Posts	?	?	Picnic sites		
Public lavatories			Nature trails and similar guided routes		
Isolated hotels (outside settlements)			First Aid Posts		
Youth Hostels			Lifesaving equipment (other than at First Aid Posts)	+	+
Jetties and landing facilities					